Contents

Introduction to Electricity.

Objectives:

The two fundamental electrical variables in the field of electricity are current and voltage.

No matter what electrical or electronic device you use, you should always have these two variables in mind. To understand current and voltage, it is important firstly to explain the basic structure of the atom and then introduce the concept of charge.

Finally, a simple explanation of static and dynamic electricity is given.

I- Structure of matter:

The matter around us (solid like a rock, liquid like water, or gaseous like oxygen.) is made up of infinitely small particles called molecules.

If we try to fragment molecules, we find that they are made of a multitude of small balls called <u>atoms</u> (from the Greek Atoms; which cannot be divided).

The diameter of these atoms is surprisingly small. To give an idea of their average diameter, let's say it is about 10 millionth of a millimeter! We can imagine that on a length of 1 millimeter, we could align one against the other 2 to 10 million atoms! The atom is therefore the smallest part of a simple body.

I-I- Structure of the atom :

The atom consists of a nucleus and one or more layers of electrons that revolve around it. The nucleus, as you can see in figure 1, is itself made up of neutrons and protons.

The protons carry a positive charge while the neutrons are, as their name indicates, electrically neutral, in other words, they do not carry an electrical charge.

The electron carries a negative charge equal in size to the positive charge of the proton.

An atom, in its natural state, is electrically neutral (neither positive nor negative).

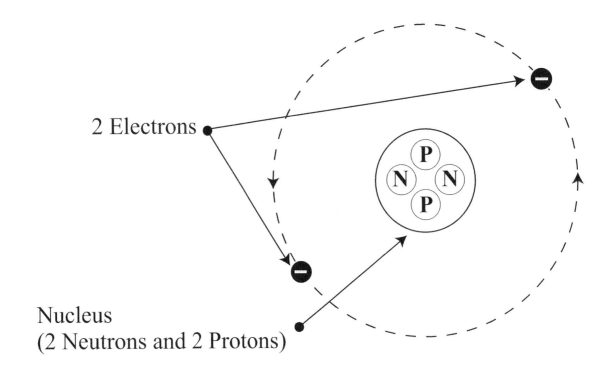

2 Electrons

Nucleus
(2 Neutrons and 2 Protons)

Figure 1: **Structure of an atom.**

To be known:

The structure of an atom has often been compared to the solar system of the universe.

The electrons would represent the planets revolving around a nucleus which would be the sun, from which the comparison between the planetary orbits and the electronic orbits of an atom.

A neutral atom contains Z electrons, Z is the atomic number; it is also the number of protons in the nucleus, which contains A nucleons; A is the atomic mass.

There are N neutrons with N=A-Z,

    A   atomic mass = the number of nucleons in the nucleus

    Z   atomic number = the number of protons in the nucleus

    N   neutron number = the number of neutrons in the nucleus

Example:

In the figure (2), you can see the distribution of electrons in the copper atom in four layers.

The electrons in the inner layers (1st, 2nd and 3rd layers) are called bound electrons. Those of the outer layers (4th layers) are called free electrons.

Copper atom: Cu: $Z = 29$ electrons so 29 protons;

A= 64 so (64-29) = 35 neutrons;

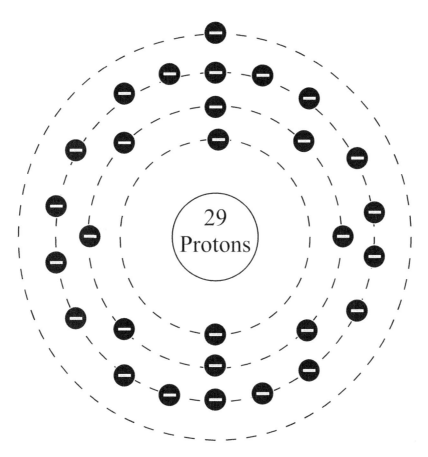

Figure 2: **Copper atom.**

Another concept to remember:

The farthest peripheral layer from the nucleus gives the atom its electrical properties.

II- Electrical Charge:

Rather than counting the number of electrons in a conductor to express the corresponding amount of electricity, a unit of electricity amount or charge called the Coulomb is defined. There are two types of charge: positive and negative.

The charge is designated by the symbol Q, and measured in Coulomb (C). The charge of an electron has the same absolute size of that of a proton. It is equal to 1.6×10^{-19} Coulomb.

Terms used: Electric charge.

Notation: Q

Unit : Coulomb

Symbol : C

6 Protons	6 Protons	6 Protons
6 Electrons	8 Electrons	4 Electrons
This atom is Neutral	This atom is Negatively Charged	This atom is Positively Charged

III- Definition of electricity:

Electricity is the movement of free electrons in a conductor. In fact, free electrons are attracted to atoms that have lost electrons and therefore to electrically positive atoms.

Normally, the movement of free electrons tends to be balanced in matter. In an electric circuit, the current source causes billions of electrons to move in the same direction. Electricity is therefore an ordered movement of free electrons in matter.

This electron will only be released, and therefore available for conductivity, if it receives an energy higher than the one that binds it to the atom. This energy can be in various forms:

1- Calorific or thermal form (thermoelectric batteries, electronic emission from filaments);

2 - Luminous form (photoelectric cells);

3 - Magnetic and mechanical form (case of electromagnetic generators):

      - Dynamos for direct current;

      - Alternators for alternating current.

4 - Chemical form (batteries).

IV- Static electricity:

Static electricity is produced by friction or contact of two nonmetallic materials and causing a simple grouping of electrons.

- A plastic rod rubbed with silk attracts light bodies ➡ the plastic rod becomes a charged body.

- A glass rod rubbed with wool also attracts light bodies ➡ the glass rod becomes a charged body.

When a body becomes charged, either positively or negatively, it is said to have static electricity. In other words, static electricity is electricity whose negative and positive charges are immobile.

- Two frictionally charged glass or plastic rods repel each other.

- A charged plastic rod attracts a glass rod and vice versa.

Static discharge can cause damage to electronic components if proper precautions are not taken.

The figure below shows the phenomena of repulsion and attraction:

▶ Charges of the same signs (positive - positive or negative - negative) repel (a and b).

▶ Charges of opposite signs (positive - negative) attract (c).

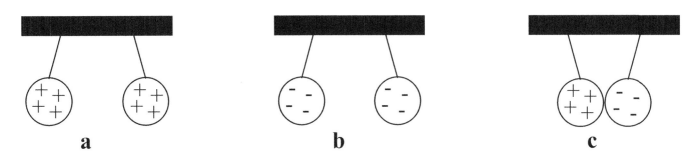

Figure 3: **Repulsion and attraction of electric charges.**

V. Dynamic electricity

Electricity in the rest state (static electricity or even electrostatic electricity) in all bodies can give no sign of its presence unless it is set in motion. Thus, by introducing an extra electron into a circuit from any source, the movement of electrons will occur as a result of a chain reaction.

Each atom reacts in such a way that it remains neutral, in other words, it always has an equal number of electrons and protons.

The movement of electrons in a conductor is similar to that of balls placed in a channel, and that one would introduce an additional one at one end of the channel moving the opposite one outside.

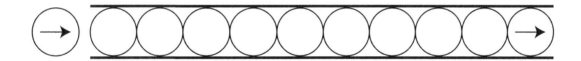

This reaction of electrons is called dynamic electricity. That means dynamic electricity, is electricity in which the electrons are constantly moving in a circuit.

To create a movement of electrons in a circuit, there must be a device capable of moving electrons and recovering them. This device is called current source.

To sum up:

1. An atom contains three kinds of elementary particles:

       - The electron of negative charge;

       - The positively charged proton;

       - The uncharged neutron.

2. The elements differ from each other by their atomic number Z.

3. Two charged bodies of opposite sign attract each other.

4. Two charged bodies of the same sign repel each other.

5. Static electricity means immovable free electrons.

6. Dynamic electricity means free electrons in motion.

Classification of Materials.

Objectives:

Firstly, we will classify materials according to their resistivity. Then, we will present the settings influencing the resistance of these different materials.

After a careful study of this lesson, you will be able to easily:

- Classify materials into conductors, semiconductors and insulating materials.

- Know the essential settings that change the electrical resistance of conducting, semi conducting and insulating materials.

I- Classification of materials:

Since materials do not all have the same atomic structure and the same number of free electrons, they therefore have different resistance and resistivity values. The nature of a material is consequently characterized by a factor called resistivity.

I-I- Resistivity of materials:

The resistivity reflects the ease with which, in a given material, an electron can be torn from its orbit around its nucleus. For each material and each temperature the resistivity is constant.

Terms used: Resistivity;

Notation: ρ (rho);

Unit: Ohm-meter;   Symbol : $\Omega.m$.

The resistivity ρ of any material is a constant characteristic of that material. Some resistivity values of common materials:

Materials	Resistivity expressed in power of 10
Silver	1.6×10^{-8} $\Omega.m$
Copper	1.7×10^{-8} $\Omega.m$
Gold	2.2×10^{-8} $\Omega.m$
Aluminum	2.8×10^{-8} $\Omega.m$
Tungsten	5×10^{-8} $\Omega.m$
Iron	8.5×10^{-8} $\Omega.m$

We will therefore define, depending on the resistivity several classes of materials.

I-II- Insulating materials:

These are substances whose atoms have a saturated peripheral layer. In other words, electrons cannot pass from one atom to another. So the resistivity is very high, we cannot tear electrons to the nucleus and therefore, no electron, no current, the material is insulating.

Insulating materials used in electronics:

This materials are used when we want to prevent the passage of a current. Among the insulating materials used in electronics, we find:

Glass, ceramic, mica, paper, porcelain, rubber, ebonite ...etc.

I-III- Electrical conductors:

Not all materials are equally opposed regarding the movement of electrons. In some materials, peripheral electrons are held so weakly by their atoms which jump spontaneously from one atom to another. The conductors of electricity will therefore be the materials exchanging electrons between their atoms. That is to say, bodies whose peripheral layer can give or accept one or more electrons.

The less good the conductor, the more work is required to draw electrons from their atoms; it is said that this conductor opposes certain resistance to the circulation of electrons.

Thanks to the conductive wires, the connection between the different components of an electric circuit (generators, receivers ...etc.) is ensured.

Conductive materials used in, electronics:

Most metals are good conductors. Silver and copper are the best conductors because they have many free electrons.

For applications requiring low resistivity, light mass and good mechanical holding, typical for antennas, aluminum and/or its alloys are chosen.

For applications requiring good contacts, gold is chosen but its cost limits its use.

To be known:

An electrical cable is the combination of several conductors mechanically joined by the outer covering but insulated from each other.

Cables and electrical conductors are defined by a code engraved on the outer envelope. This code designates the nominal voltage, the shape and flexibility of the cable, the nature of the insulation and the protective sheaths.

I-IV- Electrical semi-conductors:

The properties of these materials are common with those of both insulators and conductors. Let us explain briefly this paradox.

In its pure state, a semiconductor, placed in conditions of very low temperatures (0°absolute, equal - 273° centigrade) will have all the properties of an insulator. In fact, in these circumstances, all the electrons of the semiconductor are bound ... or if we prefer, the conductivity is zero.

Reminder

The two most commonly used units for temperature are degrees Celsius and degrees Kelvin. The mathematical relationship between the two units is :

$$T\ (°K) = T\ (°C) + 273$$

On the other hand, if the temperature increases, some electrons acquire sufficient energy to be released and ensure a certain conductivity. This latter will also be favored if we incorporate some foreign atoms in the semiconductor crystal (operation called Doping of semiconductors).

It should be noted that thermal energy has a great influence on the behavior of a semiconductor material.

In summary, a semiconductor becomes a conductor if it is put in favorable conditions (sufficient temperature, doping ...etc.).

Semiconductor materials used in electronics:

In the category of semiconductors, we will mention Silicon (Si) and Germanium (Ge). These are the two most widely used semiconductors, as they play an essential role in the conception of modern electronic devices such as diodes, transistors and thyristors.

II- Settings influencing the resistance of materials:

It should be noted that the resistance of a conductor depends on four factors:

- The temperature T.
- The length of the wire L;
- The wire section S;
- The type of materials, that is its resistivity ρ.

II-I- Influence of the temperature on the resistance:

Resistance has temperature coefficients. This means that its value changes according to temperature variations. If this effect is not desired, it is harmful, which is easy to conceive. Some kinds of particular resistance exploit this phenomenon. These coefficients can be positive or negative.

There is a formula that allows the calculation of the resistance value at any temperature knowing its value at 0°C and its temperature coefficient:

$$R(T) = R(T_0)(1 + \alpha.T)$$

With:

R(T): Value of the resistance at temperature T in °C.

$R(T_0)$: Value of the resistance at temperature 0°C.

α: Temperature coefficient.

T: Temperature in °C reached.

II-II- Influence of the length on the resistance:

Let's consider a wire of constant circular section S and length AD.

Let us divide this length "AD" into three unequal portion (see diagram below).

The materials AB, BC and CD are analogous.

$L_1 = AB = 4m$

$L_2 = BC = 8m$

$L_3 = CD = 12m$

These 3 portions welded in a row are connected to three identical multimeter allowing to measure the resistance of each section.

M_1 to the terminals of L_1

M_2 to the terminals of L_2

M_3 to the terminals of L_3

Let's read the indications of M_1, M_2 and M_3:

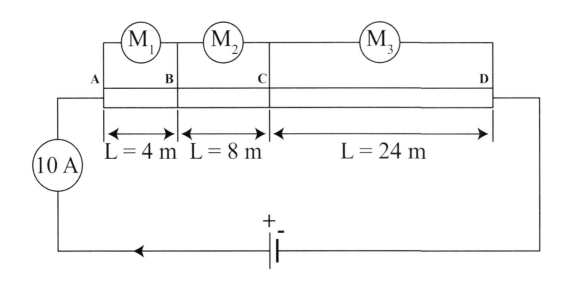

Let's record the indications of the measuring devices on a table called the measurement table.

Length	Multimeter	The measured resistance R
L_1	M_1	$R_{AB} = 2\ \Omega$
L_2	M_2	$R_{BC} = 4\ \Omega$
L_3	M_3	$R_{CD} = 6\ \Omega$

We notice that:

The resistance increases, If the length increases.

$$R = 2\ \Omega \longrightarrow L = 4\ m$$

$$R = 4\ \Omega \longrightarrow L = 8\ m$$

$$R = 6\ \Omega \longrightarrow L = 12\ m$$

To remember:

The resistance is directly proportional to the length of the conductor.

Hydraulic analogy:

Let's consider, the two following diagrams; each system is composed of two tanks connected by a pipe of length L_1 for the first and L_2 for the length with L_2 greater than L_1.

At the beginning, the valve R being closed, the difference in water level between the two tanks is identical for both systems.

At the beginning, the valve R being closed, the difference in water level between the two tanks is identical for both systems.

If we open both valves at the same time, we can see that the upper tank empties into the inner tank in a time t_1 for the first system and t_2 for the second with t_2 slightly higher in t_1.

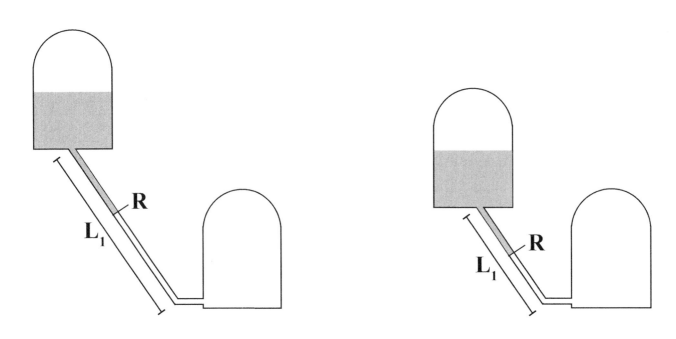

We note that:

The time of passage of the water flow is longer than the length pipe. So, the water flow, which means the quantity of water flowing per second is more important as the pipe is short.

We will say that the pipe AB is less resistant to the passage of water than the pipe BC.

II-III- Influence of the section on the resistance:

The assembly below is composed of three conductors AB, BC and CD, of the same length and respective section S1, S2 and S3.

The sections S1, S2 and S3 welded together are of the same length and nature, but of different sections.

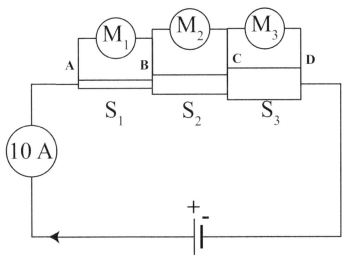

The multimeters M1, M2 and M3 connected in parallel to the good parts of AB BC CD have the same characteristics.

Here again, as in the previous paragraph, let's take the measurements in table form:

Section	Multimeter	The resistance R measured
S_1	M_1	$R_{AB} = 4\ \Omega$
S_2	M_2	$R_{BC} = 2\ \Omega$
S_3	M_3	$R_{CD} = 1\ \Omega$

We note that

The resistance decreases, If the section increases.

$$R = 4\,\Omega \longrightarrow S = 0.3 \text{ inch}^2$$

$$R = 2\,\Omega \longrightarrow S = 0.6 \text{ inch}^2$$

$$R = 1\,\Omega \longrightarrow S = 0.9 \text{ inch}^2$$

The continuous electrons in the reserve V will pass more easily in the conductor BC, because the section is greater than that of AB.

Consequently, BC has a resistance, lower than that of the conductor AB this will be written: $R_{BC} < R_{AB}$.

To remember

The resistance of an electrical conductor is inversely proportional to the section of the conductor.

II-IV- Influence of the resistivity on the resistance :

Let's imagine 3 portions of conductors AB, BC and CD of the same length L, of the same section S but of different nature (different resistivity):

Exemple:

AB: Copper $\longrightarrow$ resistivity ρ_1

BC :   Aluminum $\longrightarrow$ resistivity ρ_2

CD :   Lead $\longrightarrow$ resistivity ρ_3

The three wires (AB, BC and CD) being in a series, the generator will supply them simultaneously with the same intensity.

Here again, as in the previous paragraph, let's take the measurements in table form:

Materials	Multimeter	The resistance R measured.
ρ_1	M_1	$R_{AB} = 2.5\ \Omega$
ρ_2	M_2	$R_{BC} = 4\ \Omega$
ρ_3	M_3	$R_{CD} = 32.5\ \Omega$

For length and analogous section, the resistance of lead wire is greater than that of aluminum wire, which is greater than that of copper wire.

$R = 2.5\ \Omega$ Copper;

$R = 2.5\ \Omega$ Aluminum;

$R = 2.5\ \Omega$ Lead.

The resistance R of a homogeneous conductor is therefore a function of this conductor nature.

It is directly proportional to the resistivity of the conductor p.

After this study, we can deduce the following formula of the resistance R for a wire of **length L(m)**, of **transverse surface S(m²)** and of **resistivity** $\rho(\Omega.m)$:

$$R = \frac{\rho L}{S}\ (\Omega)$$

This formula shows that the resistance is :

- ▶ Proportional to the length of the wire L.
- ▶ Inversely proportional to the section of the wire S.
- ▶ Dependent on the intrinsic resistivity of the material ρ.

To remember

1. Since the number of free electrons and their distance from the nucleus varies for each material, it is also a factor of resistance, each material therefore has a specific resistance.

2. The smaller the diameter of a conductor, the more difficult it is to pass a certain number of electrons, so the resistance is greater.

3. Raising the temperature of a material causes the increasement of electrons' movement velocity in that material, which generally increases the number of electrons collisions with atoms.

The more collisions there are, the greater the resistance is.

4. Resistance also increases as the length of a conductor increases. The longer the path to travel, the more collisions there are within the conductor itself and hence the higher the resistance.

Bottom line:

Materials are classified according to their resistivity into three categories:
- **Insulators**: materials that have a high resistivity factor so they are poor conductors of electricity, example: glass.
- **Conductors**: materials that have a low resistivity factor, so they are good conductors of electricity, e.g. copper Cu.
- **Semiconductors**: materials that have a resistivity factor between insulators and conductors depending on the conditions where the semiconductor material is located, e.g. silicon Si.

The electrical resistance of a material depends on four basic factors:
- **The temperature of the material T**: for many materials, the higher the temperature, the greater the resistance;
- **The section S: the larger the section**: the lower the resistance;- the length of the conductor L: the longer the, conductor, the higher the resistance.
- **The length of the conductor L**: the longer the conductor, the higher the resistance.
- **The nature of the material**, i.e. its resistivity ρ: some materials offer more resistance than others;

Electrical Current

Objectives:

The origin, direction and intensity of electric current are essential notions to understand the different effects of electric current on prevention and maintenance. Two types of current are commonly used: direct and alternating current. Don't worry, this lesson will explain all these concepts in a simple and effective way.

After careful study of this lesson, you will easily be able to:

- Know the origin, direction and intensity of electric current;

- Discriminate between direct and alternating current;

- Know the effects of the electric current passage.

I. Origin of the electric current:

The movement of electrons from one part of matter to another part of matter is called electric current. The electric current is therefore the quantity of electrons that circulate in a conductor. The higher the electric current, the more electrons are circulating.

How will the electrons move?

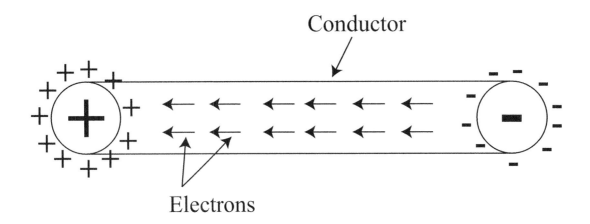

If a voltage is applied across a conducting or semiconducting material, one end becomes positive and the other becomes negative (Figure Bl). The repulsive force due to the negative charge on the right pushes the electrons (negative charges) to the left. The attractive force of the positive charge on the left attracts the free electrons to the left.

The resulting motion is a wave of free electrons going from the negative end to the positive end. This is called the electric current. Its symbol is I and its unit is the ampere (A).

We can also say that: the current is the rate of the charge flow.

II. Direction of the electric current:

First of all, it is necessary to differentiate between the conventional direction and the real direction or electronic direction. When the current circulates, in an electric circuit, from the positive terminal (+) towards the negative terminal (-), it is what we call the conventional direction of the electric current.

According to electronic theory, electric current always flows from a negative charge (-) to a positive charge (+). The positive charges attract the negative electrons in order to balance the atoms. Consequently, in an electric circuit, the current flows from the negative (-) terminal to the positive (+) terminal. This is called the real or electronic direction of the electric current.

III. Electric current intensity:

It is essential to measure the amount of current flowing in an electrical circuit. The basic unit used to do this is the coulomb. One coulomb represents about 6.24 billion electrons. By counting the number of coulombs passing through a conductor in a given time, we can determine the intensity of the electric current.

The SI (International System of Units) symbol for ampere is I.

To measure the number of electrons passing through a conductor, we use a device called an ammeter or another device called a multimeter.

Multimeter

35

Mathematical formula:

The current in a conductive material is calculated by the number of electrons (amount of charge) that crosses a point per unit of time.

Hence the fundamental relationship:

$$I = \frac{Q}{t}$$

Terms used: Current intensity, ampere

Notation: I

Unit: Ampere

Symbol : A

To know:

An Ampere (1 A) is the amount of current that exists when a charge of one coulomb (1 C) crosses the cross-section of a conductor in one second (1s).

For a current of 2 amperes, we will note: I = 2 A

For a current of 0,1 ampere, we note: I = 0, 1 A

Exercise

1) What would be the intensity of a current corresponding to the circulation of 1800 coulombs per hour?

The formula to use is: I= Q\t;

Q = 1800] C and t = 1 hour.

Reminder

Before performing a calculation on a formula, you must always pay attention to **the units**

Here, the time t must be expressed in seconds.

t = 1 hour = 60 x 60 = 3600 seconds.

So   I = 1800\3600 = 0.5 A

2) How many coulombs are carried in 10 minutes by a current of 2 amps?

Here, the problem is a little different. We know I and t and we look for Q

I=2 A

t = 10 minutes = 600 seconds.

Q= ?

From the formula I = Q \ t, we can derive: Q = I . t

(The point between I and t means I multiplied by t; usually, we do not put this point and we write Q = I t).

Q = 2 X 600 = 1200 C.

Sub-multiples of the Ampere	Symbol
Ampere	A
Milliampere	mA
Microampere	µA
Nanoampere.	nA

IV- Types of electric current:

IV-I Strong currents:

This category includes

- The currents delivered by generators of little volume and little power such as batteries and accumulators which are of 0, 1 A to 1 A order.

- On the other hand, generators such as dynamos (in direct current) or alternators (in alternating current) can deliver currents from 1 A up to several hundred amperes.

- At the output of power plants, the intensities are even more considerable and can exceed 1,000 amperes!

IV-II- Weak currents:

In electronics, the order of electrical dimensions magnitude is much smaller; here are some examples, as an indication:

- The currents flowing through the electrodes of a diode or a transistor will usually be of the order of a few mA up to 1 A

In some cases, they will not exceed the µA (micro-ampere). This has nothing to do with the intensities encountered in industrial electricity!

- The intensities through the different stages of a TRANSMITTER will obviously be Higher than those which will circulate through a receiver or a high-fidelity system!

In radio and TV reception, we are in the field of weak currents.

IV-III- Direct current:

An electric current, in a general way, is due to an agitation and a circulation of free electrons.

If this circulation always takes place in the same direction, we will say that it is a continuous current. Let's try to compare the continuous movement of electrons to that of balls in a tube. If we tilt this tube, the balls, under the effect of their weights, start moving from the highest level (A) to the lowest level (B).

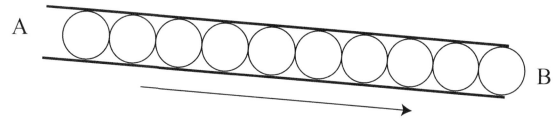

If this circulation always takes place in the same direction, we will say that it is a continuous current. Let's try to compare the continuous movement of electrons to that of balls in a tube. If we tilt this tube, the balls, under the effect of their weights, start moving from the highest level (A) to the lowest level (B).

A LEVEL DIFFERENCE is therefore required to set the balls in motion.

In electricity, a analogous condition must be respected to create the continuous movement of electrons. This difference in level is called POTENTIAL DIFFERENCE or VOLTAGE (expressed in Volt). It will be provided by the GENERATOR whose role is similar to that of a PUMP which sucks in the electrons through one of its terminals, and pushes them out through the other.

To remember

Direct current is a movement of electrons that always flows in the same direction inside a conductor.

IV-IV- Alternating current:

Let's go back to our mechanical analogy (balls in the tube) and imagine that we can change the direction of the tube's inclination at a certain rate. The balls will go sometimes towards B, sometimes towards A; thus, change direction according to the proposed inclination.

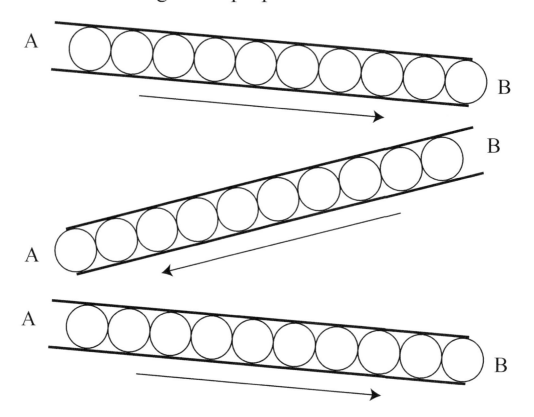

In the same way, an alternating current will result from an inversion in the direction of the electrons' movement. This inversion will be due to the generator used in such a circumstance, the generator which will be called alternator (generator of alternating current).

To remember:

The alternating current is produced by a movement of electrons which takes place alternately in both directions inside a conductor.

V- Effects of the electric current:

We have seen previously that current is a circulation of electrons in our circuits. Naturally, only the effects of electric current reveal the existence of electricity. We will detail the effects of the electric current.

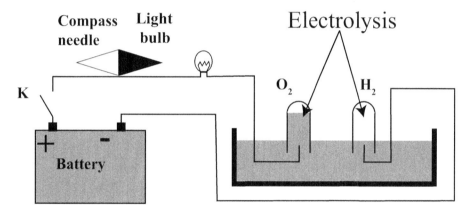

Here is the test setup:

A car battery powers a very ordinary light bulb. In series on the circuit, we find an electrolysis tank containing essentially water to which soda has been added. If we close the electrical circuit through switch K so that current flows, we find:

▶ The lamp lights up and heats up

▶ The needle of the compass has deviated

▶ A gas is formed in each tube

The, filament of the lamp, in order to emit visible radiation, is heated to a high temperature, this is the thermal effect.

The magnetic needle of the compass is moved by the magnetic field caused by the passage of the electric current, this is the magnetic effect.

The aqueous solution decomposes. Oxygen is collected on one side and hydrogen on the other. Moreover, we note a volume of hydrogen more important than that of oxygen, (H_2O) it is the chemical effect.

Polarization effects:

Let's try reversing the polarities of the battery and see what happens:

▶ The compass needle changes direction.

▶ Oxygen is now collected where hydrogen was collected and vice versa.

▶ The lamp shines with the same brilliance without any other manifestation..

Two effects out of three are polarized (they are influenced by the direction of the current flow), these are the magnetic and chemical effects, as for the thermal effect it is not polarized.

V-I- Thermal effect:

Also called joule effect, this effect is used for example in a toaster, a light bulb or a radiator. Unfortunately, this effect is not always desired and it still occurs, leading to the installation of heat dissipation systems to remove the heat. Generally, it should be noted that any overheating is synonymous with a loss of efficiency.

V-II- Chemical effect:

The most known case, electrolysis. This effect is used for the charge/discharge of batteries.

V-III- Magnetic effect:

If we place a compass near a wire carrying an electric current, we can see that the magnetized needle moves. The current has created a magnetic field. This extremely important property will be exploited extensively to produce relays, electric motors, loudspeakers, etc.

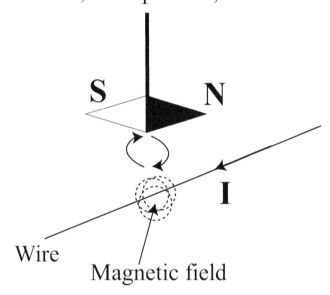

Wire

Magnetic field

To remember:

Magnetic, Mechanical and Chemical effects, depend on the direction of the current.

Thermal and luminous effects are independent of the current direction.

Electrical Voltage and Ohm's Law

Objective

The second fundamental electrical variable in the field of electricity is voltage (remember the first fundamental electrical variable in the field of electricity? ... Yes, it is the electric current, you can consult lesson 2).) Whatever electrical or electronic device you use, you should always have these two variables in mind. In this lesson, we will show you the concept of electromotive force and voltage drop, then we will move on to explain Ohm's law and through simple examples, we will demonstrate how to use Ohm's law in electrical circuits.

After careful study of this lesson, you will easily be able to:

- Explain the concept of electromotive force;
- Define voltage drop;
- Explain and practice Ohm's law.

I- Electromotive force or potential difference:

I-I- Definition:

In order that electric current flows in a conductor, the number of electrons must be different between two ends of the conductor. The greater the concentration of electrons at a given point, the greater the force of repulsion between the electrons. This force is called electromotive force (EMF). The electromotive force is therefore the force that causes the movement of electrons in matter.

The expression of potential difference, in short, is often used to express the capacity of the two charges. Positive and negative.

Let's take a simple analogy: suppose you hold a ball in balance in a tube. If you hold the tube level, the ball remains stationary; as soon as you tilt the tube, the ball starts to move. So you have created a difference in level to set the ball in motion. In electricity, it is analogous, to make electrons flow (the current) we need to produce a potential difference beforehand. The accumulation of positive charges on one side (the plus "+" pole) and the accumulation of negative charges on the other side (the minus "-" pole) creates a potential difference or voltage.

In an electrical circuit, it is the potential difference that, allows the movement of electrons through the electrical devices. The electromotive force is thus totally consumed by the electrical devices.

Good to know:

The word voltage is most often used in automotive technology to refer to electromotive force, potential difference, electrical pressure or electrical voltage.

Note:

Voltage is a measure of the energy available in each coulomb of charge.

The unit of energy amount per unit of charge is the volt.

The SI symbol for electromotive force is E.

Voltage is also designated by the symbol U or V

The instrument used to measure voltage is the voltmeter. Another instrument called the Multimeter is also used.

The symbol used to indicate the results of the measurement is V for volt.

Terms used: electric voltage, potential difference, electromotive force and voltage.

Notation: U, V or E

Unit: volt

Symbol: V

Good to know:

One volt (1 V) is the potential difference (voltage) between two points when an energy of one joule (1 J is needed to move a charge of one coulomb (1 C) from one point to the other.

Sub-multiples of volt		Symbol
Volt	1 V	V
Millivolt	10^{-3} V	mV
Micro volt	10^{-6} V	μV
Nanovolt.	10^{-9} V	nV

I-II- Hydraulic analogy:

Let us imagine a pipe P joining two tanks A and B containing water. If we assume that the two tanks are at the same level, there is no exchange of liquid between A and B.

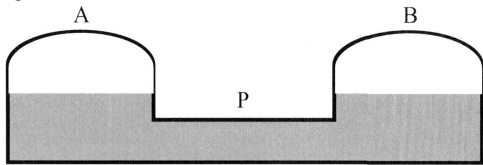

If the difference in level is zero, the flow through pipe p is zero.

To note:

No level difference <===> No flow

On the other hand, if tank A is located at a higher level than tank B, A will empty itself to the benefit of B. There is therefore an exchange or flow between A and B and a circulation of water directed from A to B, i.e. from the highest level to the lowest.

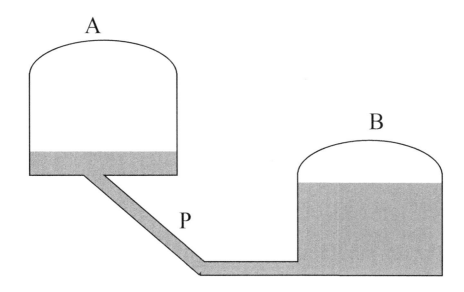

Obviously, if the level of B is higher than that of A, B will empty itself to the benefit of A, hence, this time, flow directed from B to A.

To note:

If there is a difference in level <===> There is a flow.

If level A > level B, the flow is from A to B (+ to -), as for the flow outside an electric generator.

(The sign > means: greater than ..., here: A greater than B)

How to generate a voltage?

The systems able to generate this voltage are for example: a car battery, a battery, the domestic sector, solar panels, ... etc.

II- Voltage drop:

Voltage drop is the voltage difference that exists between the two terminals of an electrical resistor or any electronic component. In other words, a voltage drop occurs when the electromotive force forces a current through an electrical resistor or any electronic component.

Good to know:

Excessive voltage drop in the headlight circuit causes low voltage to the lamps and, consequently, produces poor lighting.

III- Ohm's Law:

Here we are, the time has come to discover this famous law which is fundamental. It will guide you in all the moments of your life as a technician because it is the principle that you will apply practically everywhere, whether it is for the design of your circuits, troubleshooting or analysis. Its apparent simplicity hides treasures. It is essential to master this law.

III-I- Definition:

Ohm's law is a mathematical relation between voltage, current and resistance in an electric circuit. George Simon Ohm (1787 - 1854) established experimentally that if the voltage across a resistor increases, its current also increases.

Statement of Ohm's Law:

A voltage of one (1) volt is required to make a current of one (1) ampere flow through a resistance of one (1) ohm.

Mathematical formula:

Ohm's Law can be expressed as an equation, indicating the relation that exists between voltage U, current I, and resistance R.

$$U = R \cdot I$$

Units:

The voltage **U** expressed in **volts**, whose SI symbol is **V.** This latter is equal to the resistance **R** expressed in **Ohms**, whose SI symbol is Ω, multiplied by the current **I** expressed in **amperes**, whose SI symbol is **A.** This is an essential formula to know perfectly. It will be useful to you for solving any problem of electricity or electronics.

We will learn, while playing, and determine two important characteristics:

We are going to make the test assembly that you see on the right and proceed to measure the current and the voltage.

For the first measurement, we will keep R constant and equal to 1 Ω and we will vary U, in order to determine the current I.

R (Ω)	1	1	1	1	1
U (V)	1	2	3	4	5
I (A)	1	2	3	4	5

For the second measurement, we vary R from 1 Ω to 5 Ω while keeping U constant and equal to 10 V, in order to determine the current I.

U (V)	10	10	10	10	10
R (Ω)	1	2	3	4	5
I (A)	10	5	3.3	2.5	2

According to the figure below, the current evolves according to the applied voltage, in other words, the fact of increasing the voltage makes the current increase and vice versa.

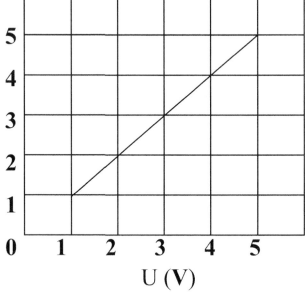

On the other hand, the following figure shows that increasing t he resistance causes the decreasement of the current.

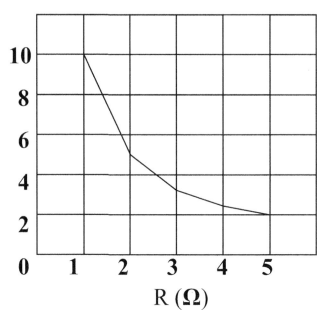

There are three equivalent forms of Ohm's law depending on the variable we want to know:

- Current formula: I = U / R

- Voltage formula: U = RI

- Current formula: R = U / I

To remember:

If we increase the potential difference, U ===> the current I increases.

If we decrease the potential difference, U ===> the current I decreases

We say that the potential difference, U, is directly proportional to the current I, or that U is a function of I.

In mathematics, this is written:

$$U = f(I)$$

III-II- Applications:

In this section we will present examples of Ohm's law applications. In these examples you will learn how to determine voltage, current and electrical resistance by analytical and graphical methods.

Analytical method:

Example 1:

What is the current intensity of the following figure:

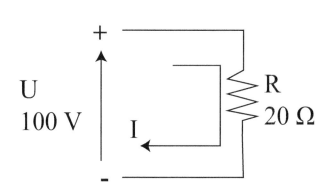

Solution:

Let's take the formula I = U/R and replace U by 100 V and R by 20 Ω.

I= U/R= 100/20 = 5A.

Reminder

In the metric system, large resistances are indicated with prefixes like kilo (k) or mega (M). Thus, instead of saying one thousand Ohms, we say one kilo-Ohm **kΩ**, and instead of saying one million Ohms, we say mega-Ohm **MΩ**.

Example 2:

What is the voltage across the resistance in the following figure:

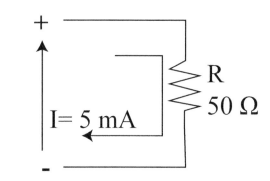

Solution:

Note that 5 mA equals 5×10^{-3} A.

Substitute the values for I and R in the formula U = RI.

$$U = RI = (5 \text{ mA}) \times (50 \text{ Ω}) = (5 \times 10^{-3} \text{ A}) \times (50 \text{ Ω})$$

$$= 25 \times 10^{-2} \text{ V} = 250\text{mV}$$

Example 3:

In the following circuit, what is the value of R so that the current drawn, from the source is 5 mA?

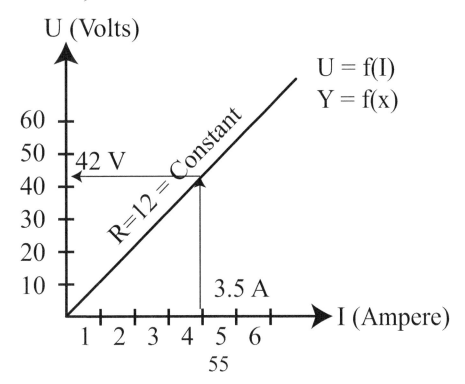

Solution:

Let's substitute 150 V for U and 5 mA for I in the formula: $R = U/I$.

$R = U/I = 150/(5 \times 10^{-3}) = 30 \times 10^{3}$ ohm.

Graphical method:

Example:

The pure resistance of a circuit is equal to 12 Ω. Study the variations of the voltage U at the terminals of this resistance when the intensity of the current I, which crosses it, varies from 0 to 5.

U (Volts)

$U = f(I)$
$Y = f(x)$

60

50 ⊤ 42 V

Constant

40

30   R=12 =

20

10   3.5 A

1 2 3 4 5 6   I (Ampere)

Solution:

U = RI or U = 12.I

This relation is of the form

Y = 12 . x

Table of variations:

X I (A)	0	1	2	3	4	5
Y U (V)	0	12	24	36	48	60

Use of the graph :

If we want to know the voltage for a current of 3.5 amperes, we can see right away that U = 42 volts.

IV- Measuring the voltage or POTENTIAL DIFFERENCE:

A potential difference is measured with a voltmeter or a multimeter.

A voltmeter is connected in parallel to the terminals of the component whose voltage is to be measured.

The + terminal of the voltmeter is connected to the point where the current I enters R. The - terminal is at the other end.

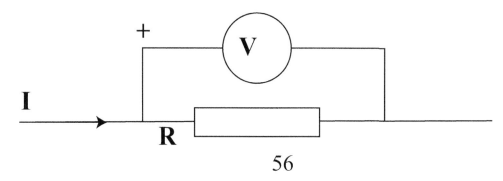

56

Essential quality of a voltmeter:

To be very resistant, so as not to consume energy and not to modify the current through R as soon as V is connected (the resistance of the voltmeter must be at least 100 times greater then that of R). On the other hand, the internal resistance of an ammeter must be low so as not to introduce too much voltage drop in the circuit.

According to the importance of the studied voltage, we can use the multiples or the sub-multiples of the volt.

MULTIPLES OF THE VOLT	SUB-MULTIPLES OF VOLT
Kilovolt= 1000 volts KW shortly Field of application: radio and television transmitters.	Millivolt = 0.001 volt mV shortly The millivolt is the order of magnitude of the voltage between the base and the emitter of a transistor.
Megavolt= 1000000 volts MV shortly Field of application: industrial electricity, power plants, high voltage lines.	Microvolt = 0.000001 volts µV shortly Field of application: The H.F. or radio waves captured by a receiving antenna. They have the order voltage of a few microvolts.

To sum up:

- Ohm's law defines a precise relation between the applied voltage, the circuit resistance and the current intensity. It is stated as follows:

- In a portion of circuit " AB " where the consumed electric energy is completely transformed into calorific energy (pure resistance R), the potential difference or voltage U is equal to the product of its resistance R by the current I

- A voltage of one (1) volt is necessary to make circulate an intensity of one (1) Ampere through a resistance of one (1) ohm

- Knowing two of these three data, it is possible to find the third one with a simple calculation. Thus, from a mathematical formula, we can determine that in an electric circuit:

- The current is directly proportional to the applied voltage;

- The resistance is inversely proportional to the intensity.

- If we want to increase the current in an electric circuit, we can either increase the voltage or reduce the resistance.

- If we want to decrease the current in an electric circuit, we can either reduce the voltage or increase the resistance.

Electrical Power and Joule's Law.

Objectives:

Still in the fundamentals of electricity and electronics, in this fifth lesson we will introduce the concepts and definitions of energy and power as well as Joule's law in electrical circuits.

After careful study of this lesson, you will easily be able to:

- Define energy and power;
- Know the relation between power and energy;
- Calculate power in an electrical circuit;
- Know the Joule effect and its applications.

I- Energy:

It is time to tackle this topic. Now, we know the main constituents of electricity. Firstly, we will go back to the notion of energy.

Definition:

Let's suppose that you have to transport a pile of soil from your garden from point A to point B with a wheelbarrow.

The contents of your wheelbarrow will be a load that you will move, to make the wheelbarrow roll you will need energy. When you have completed your mission after a few hours, you will be happy to see your work. So, energy is the ability to do a job.

It is the same in electricity. The charges (the electrons) moved by the action of the voltage **U** in a time t carry out a work, therefore they need energy W.

Terms used: Electrical energy.

Notation: W

Unit: Joule

Symbol: J

This energy can be quantified by the formula:

$$W = \textbf{U.I.t}$$

With:

W in joules (J);   U in Volts (V);   I in Amps (A);   t in seconds (s).

II- Electrical power :

Definition:

The electrical power P is a measure of the amount of electrical energy converted in one second into another form of energy by any electrical device. In other words it is the absorption or release of a certain amount of energy per unit of time (1 second), which gives:

$$P = w/t$$

And as:

$$W = U.I.t$$

The formula becomes:

$$P = U.I.t / t = U.I.$$

Terms used: Electrical power.

Notation: P

Unit: Watt

Symbol: W

The formula to calculate the electrical power is therefore:

With:

P in **Watt** symbol **W**;

U in **Volt**; $\implies \boxed{P = U I}$

I in **Ampere.**

Note:

The *W* energy symbol is italicized, while the right W is the watt symbol.

Good to know:

The watt is the power dissipated across a portion of a circuit when the potential difference is 1 volt and the current is 1 ampere.

II-I- Power in an electric circuit:

An example of power calculation:

Let's take an example that will make us review generators as well.

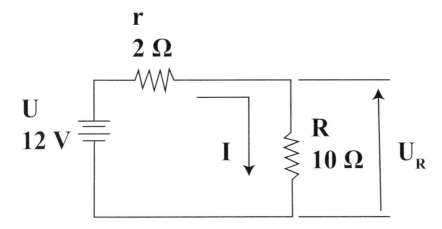

We want to know the power consumed by the resistance R, the generator having an internal resistance of 2 Ω.

To calculate this power, we need to know the voltage (UR) to the terminals of this charge resistance (R) and the current that crosses it.

1- Let's calculate the current in the circuit:

$$I = \frac{U}{(R+r)} = \frac{12}{(10+2)} = \frac{12}{12} = 1A$$

2- Let's calculate the voltage across **R**:

$$U_R = R . I \qquad ==> \qquad U = 10 \times 1 = 10\ V$$

3 - Now let's calculate the power absorbed by R:

$$P = U_R . I \qquad \Longrightarrow \qquad P = 10 \times 1 = \mathbf{10\ W}$$

If we are interested in the power supplied by the generator now.

We know that: $I = 1$ A and $U = 12$ V,

It comes:

$$\mathbf{P = U . I} \Longrightarrow P = 12 \times 1 = \mathbf{12\ W}$$

12 W are supplied by the generator, 10 W are absorbed by the charge and the 2 that are missing?

They are consumed in pure loss (heat) in the internal resistance of the generator.

To be remembered:

Any overheating, when this effect is not desired, is synonymous with loss of efficiency and lost energy.

Other useful formulas for calculating power :

The formula $\mathbf{P = U . I}$ is not the only one to determine the power, we can use the resistance to make this calculation.

$$\mathbf{P = U^2 / R}$$

$$\mathbf{P = R . I^2}$$

Let's take the previous example and calculate the power absorbed by R:

1st case:

$$I = \frac{U^2_R}{R^2} \quad ; \quad P = \frac{10^2}{10} = \frac{100}{10} = 10 \text{ W}$$

2nd case:

$$P = R.I^2 \; ;$$

$$P = 10 \times 1^2 = 10 \text{ W}.$$

The Kilowatt-hour (kWh), a unit of energy:

According to the following formula $W = P. t$ and since power is in watts and time is in seconds, the unit of energy can be referred to as **watt seconds (Ws)**, **watt hours (Wh)** or **kilowatt hours (kWh)**.

When you pay your electricity bill, the rate is based on the amount of energy consumed. The most convenient unit is **kWh.**

Example:

Determine the number of kilowatt-hours consumed by the following activities:

a) 1400 W in 1 hour,

b) 2500 W in 2 h,

Solution:

a) 1400 W = 1.4 kW

$W = P.t = (1.4 \text{ kW})(1 \text{ h}) = 1.4 \text{ kWh}$

b) 2500 W = 2.5 kW

$W = P.t = (2.5 \text{ kW})(2 \text{ h}) = 5 \text{ kWh}$

II-II- Rated power of resistance:

Recall that a resistance gives off heat when a current flows through it. There is a limit to the amount of heat a resistance can dissipate. It is determined by its rated power.

Rated power is the maximum power that a resistance can dissipate without being damaged by excessive heat accumulation, .

Carbon resistances are sold in the market with standard rated power: 1/8 W, ¼ W, ½ W, 1 W and 2 W.

Choosing the right rated power for an application:

When a resistance is placed in a circuit, its rated power must exceed the maximum power that will be applied to it. For example, if a carbon resistance is to dissipate 0.75 W in a circuit, its power rating should be the first standard value that exceeds this maximum, i.e. 1 W.

Example:

Choose the correct rated power for the carbon resistance in the circuit shown:

(1/8 W, ¼ W, ½ W, 1 W and 2 W).

10 V

R
62 Ω

Solution:

The actual rated power is:

$$P = \frac{V^2}{R} = \frac{(10\ V)^2}{62\ \Omega} = 1.6\ W$$

Therefore, select a resistance with a rated power that exceeds the actual power, in this case, we will select **a 2 W resistor.**

Damage to resistances:

When a power applied to a resistance exceeds its rated power, it will undergo excessive heating. The result is a melting of the part or a significant alteration of its ohmic value.

Example:

Determine if the resistance in the circuit of the opposite figure would not be damaged by overheating.

9 V

¼ W
100 Ω

Solution:

The actual power applied to the resistance is:

$$P = \frac{V^2}{R} = \frac{(9V)^2}{100\Omega} = 0.81 \text{ W}$$

The resistance is rated at ¼ W (0.25 W), which is insufficient to handle the applied power. The resistance was overheated and probably burned out or opened.

III. Joule's Law:

We have already discussed, without specifying it, this topic with the power formula $P = RI^2$. Let's look at this phenomenon with more details.

Definition:

The Joule effect is the heating phenomenon that appears in an ohmic conductor through which an electric current flow.

The electrical energy lost by Joule effect in a load is proportional to the time (t), to the square of the intensity (I^2) and to the electrical resistance of the conductor (R)

$$\boxed{W = R\,I^2\,t}$$

III-I- Applications:

Among the many applications originated from this phenomenon we can cite:

Incandescent lamps:

The light bulb consists of a glass bulb, inside which is located a tungsten filament. This material has a very high melting temperature, (about

3000°C), which allows it to incandescent at about 2500°C. To obtain an intense luminous effect, a rare gas is incorporated in the bulb (krypton or argon). We avoid any trace of oxygen which would burn the fila-ment. On the base of this tube, we specify the electrical voltage of using in volt as well as the power of lighting (example: 25 W, 60 W, 75 W, W being the abbreviation of the unit of power, i.e. the WATT); see at the end of this lesson.

Tungsten filament

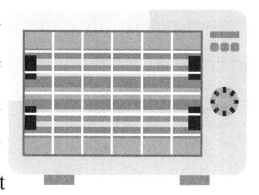

Electric heating:

This is a very flexible heat source in terms of adjustment, free of any pollu-tion. The system is made up of alloy resis-tance (ferronickel for example) which give off heat to the surrounding environment when the electric current passes through them. The resistance are wound on a refractory support (e.g. porcelain, ceramic). The heating mode can be regulated with thermo-stats.

Household appliances:

This is the case of electric kettles, iron, soldering iron, water heaters, etc.

Fuses:

They protect an installation against possible over currents due to:

- Over-power due to an excess of used receivers;

- A clumsiness in a connection;

- An insulation fault between two conductors causing a "short circuit".

To limit the current, we will place a fuse or "circuit breaker" in the circuit, at the entrance of the installation. If the current rises, the fuse reaches its melting point; the circuit is cut off and the current no longer flows.

The diameter or " caliber " of the fuse is calculated according to the maximum intensity that a given circuit can support (the materials used are lead, tin or silver alloys which have rather low melting temperatures (lead 327° C, tin 231° C. silver 960 °C).

We have practically reviewed the applications of the Joule effect, i.e., its numerous ADVANTAGES. But the Joule effect does not have only advantageous aspects.

Disadvantages:

- In electronic circuits, the resistance used tend to heat up under the passage of current. The rise in temperature modifies the value of these resistances. The result is an unstable operation due to the variation of the circuit components. The problem of heating resistances should be studied carefully when building an electronic circuit.

- The problem is even more acute in semiconductors (diodes, transistors etc.). The electrode that collects the electricity-carrying elements (the collector for a transistor) is subjected to INTENSIVE BOMBING.

The result is excessive heating and the need to predict the rejection of this "excess heat"; cooling can be done in several ways such as radiators.

The use of a cooling radiator; the transistor or the diode is integral with the radiator and its fins, in contact with the ambient air, they evacuate the thermal surplus. When the power of the transistor does not exceed a few watts, this process remains valid.

The figure below shows an integrated circuit amplifier with a power of 100 watts. This amplifier is inserted in a metal case with 8 cooling fins.

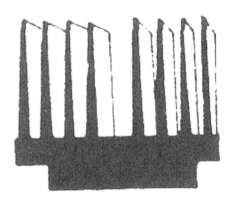

To remember:

One watt (1 W) is the amount of energy supplied during one second.

Alternating Current and Voltage.

Objectives:

We saw in the first course devoted to direct current, that it is characterized by an overall unidirectional movement of electrons. The electricity distributed by public utilities is in the form of alternating voltage and current; therefore, this second lesson represents an introduction to electrical circuits of alternating current:

Firstly, we will deal with some notions of trigonometry. Then, we will move on to the definition of the sinusoidal wave as well as its different characteristics (such as period, frequency, RMS value,…) We will end with the angular measurement.

After careful study of this lesson, you will easily be able to:.

- Define a sine wave;

- Define and determine the period and frequency of a sine wave;

- Know the different values (RMS, average, ...) of sinusoidal voltage and current;

- Know the angular measurements and conversions.

I. Importance of the sinusoidal regime:

- Most of the energy is produced as sinusoidal alternating current;

- Sinusoidal functions are simple to handle mathematically and electrically;

- Any periodic function of any form can be decomposed into a sum of sinusoidal signals.

II- Mathematical concepts:

II-I- Sine and cosine functions:

The **sine (symbol: sin)** and **cosine (symbol: cos)** functions are functions like the others, we write $f(x) = \sin x$ or $f(x) = \cos x$

Let's draw a circle of radius "r" which will always be 1.

The horizontal axis is the cosine axis.

The vertical axis is the sine axis.

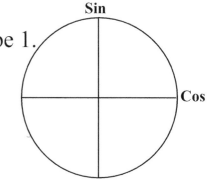

We want to determine the sine and cosine of an angle θ equals 60°. With the graphical method, nothing could be simpler.

We draw the circle of radius 1, positions it with a protractor of angle 60° and we carry out the projection of the point obtained M on the circle on the axes sine and cosine.

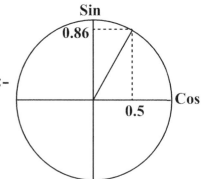

Now we take a ruler for measuring. We measure 0.5 on the cosine axis and 0.86 on the sine axis, so:

$$\text{Sin}(60) = 0.86 \qquad ; \qquad \text{Cos}(60) = 0.5$$

II-II- Measurement of angles:

The unit of measurement for angles is the degree (its symbol is: °).

A complete cycle contains 360°. One degree (1°) is an angle measurement that corresponds to 1/360° of a complete cycle or rotation.

Angles can also be measured in **radians** (the symbol is: **rad**).

A radian (1 rad) is an angular distance along the circumference of a cycle or a complete rotation, it is equivalent to 57.3°. There are 6.28 radians in a cycle of 360°.

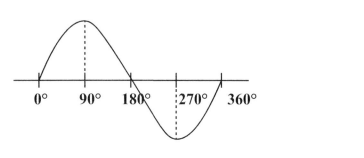

Remark

We can write: 6.28 radians = 2π, the Greek letter π (pi) represents the ratio of the circumference of any circle to its diameter, it is a constant worth 3.1416. The following table represents some values in degrees and their equivalent in radians:

Degrees (°)	0	45	90	180	360
Radians (rad)	0	π/4	π/2	π	2π

Conversion radians/degrees:

Radians are converted to degrees using the equation :

$$rad = \left(\frac{\pi \; rad}{180°}\right) x \; degree$$

Degrees are converted to radians using the equation :

$$rad = \left(\frac{180°}{\pi \; rad}\right) x \; degree$$

Example:

Convert:

a- 60 ° to radians;

b- π/6 rad to degrees.

Solution:

a-   π/3 = 1.0466 rad

b-   30°

III- Sinusoidal currents and voltages:

The sine wave is a very common type of voltage and current. Among the sources of sine waves, there are rotating electrical generators (alternators). The graphic symbol used to represent a sinusoidal voltage source is:

The figure below shows the general shape of a sine wave, which could be an alternating current or an alternating voltage. Note that starting from zero, the wave increases to a positive maximum, returns to zero and then decrease to a negative maximum before returning to zero and closing the cycle.

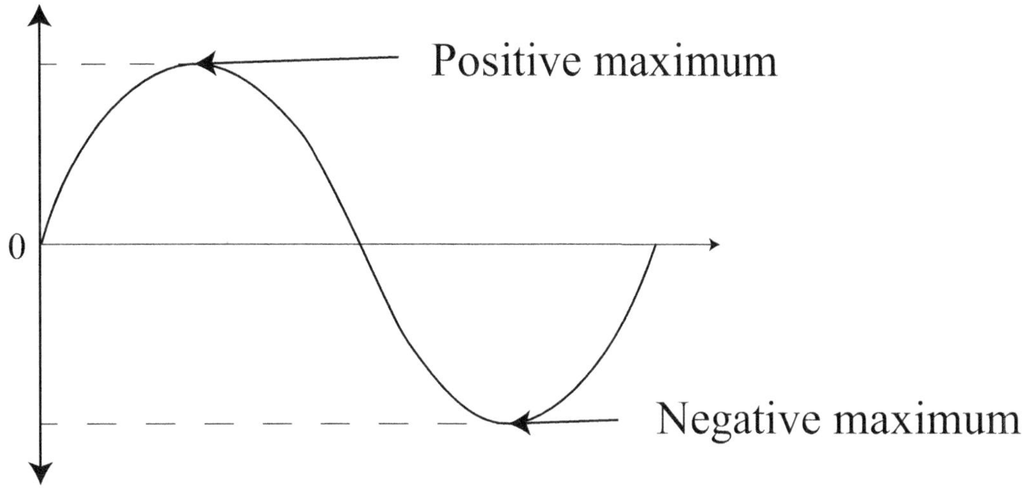

III-I- Expression of a sinusoidal voltage:

Parametric expression

Thus! A sinusoidal voltage is a periodic and alternating quantity that can be written in the form:

$$u(t) = U \sin(\omega t + \theta_u)$$

With

t is the time in seconds (s);

ω is the pulsation in radians per second (rad.s^{-1});

$\omega t + \theta_u$ is the instantaneous phase in radians (rad);

θ_u is the phase originally in radians (rad).

Positive voltage, the indicated direction of the current.

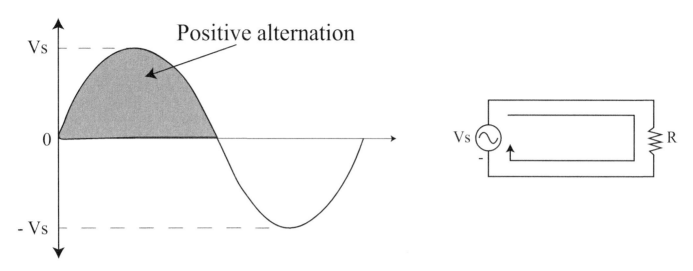

Negative voltage, the direction of the current is reversed.

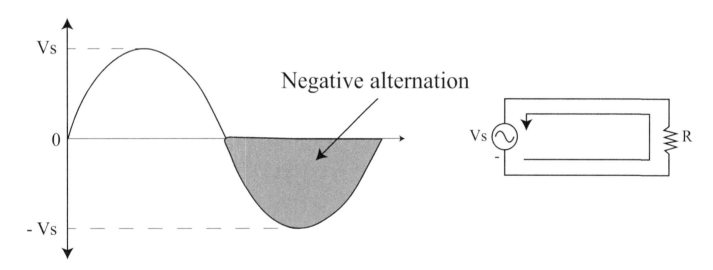

You can notice that the polarity of the sine wave changes when it goes to zero. We can say that it alternates between positive and negative values (figures above).

When a sinusoidal voltage is applied to a resistive circuit, the resulting current is also sinusoidal; and if the voltage changes its polarity, the current also changes direction. The sum of the positive and negative alternations forms a cycle of a sine wave.

III-II- Characteristic parameters of an AC voltage

III-II-I- The period: T

The time required for a sine wave to make a complete cycle is called the period and its symbol is: **T**. This period will be measured in **seconds (s)**.

The following figure represents the period of a sinusoidal wave. We can see that this wave is a periodic function because it repeats its basic cycle.

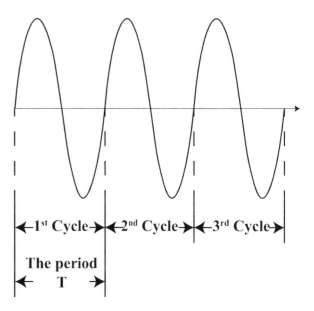

III-II-II The frequency: f

The frequency is the number of complete cycles performed by a wave in one second (1 s). The symbol of the frequency **f**. It is measured in **Hertz (Hz)**. One Hertz (1 Hz) is equivalent to one cycle per second. Thus, 50 Hz means 50 cycles per second.

Figure A shows a sine wave that makes three complete cycles per second (3 Hz), while the wave in figure B makes six cycles per second (6 Hz).

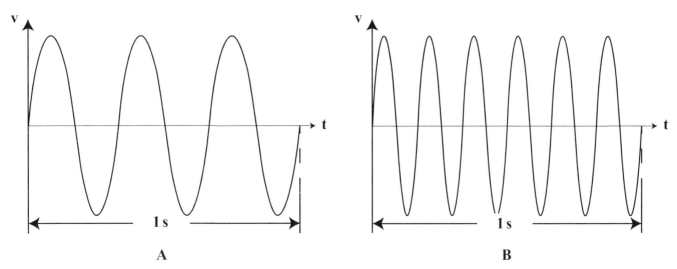

Relationship between period and frequency:

The relationship between frequency and period is very important:

$$f = 1/T$$

$$T = 1/f$$

Good to know:

In the USA or in JAPAN the frequency is 60 Hz, in aeronautics the production of embarked energy turns to 400 Hz.

In Great Britain, the grid frequency is 50 Hz.

III-II-III- The pulsation: ω

The pulsation is the angular velocity of a wave; it has the symbol: ω and it is expressed in radians/second (rad/s). We can calculate the pulsation from the frequency using the following formula:

$$\omega = 2.\pi.f$$

Reminder:

An angle θ can be expressed in degrees (°) or radians (rad). The mathematical relationship between the two units is:

$$\theta(rad) = \frac{\theta(°).\pi}{180}$$

With the symbol π is equal to 3.14

III-II-IV- Phase shift: φ (phi)

Firstly, the phase of a sine wave is an angular measure of the position of the wave relative to a reference (Figure opposite).

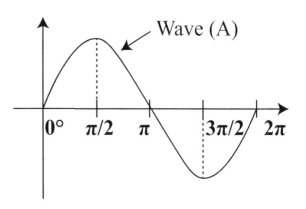

Now when we trace another wave (B) shifted to the left or right with respect to wave (A) as shown in the figures below, we say that there is a phase shift.

The wave (B) is shifted to the right with respect to the wave (A). We say that (B) is behind (A) by π/2, or (A) is ahead of (B) by π/2.

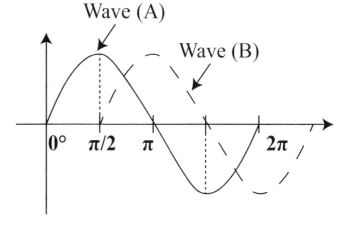

The wave (B) is shifted to the left with respect to the wave (A). We say that (B) is ahead of (A) by π/2, or (A) is behind (B) by π/2.

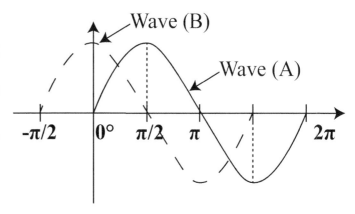

III-II-V- Phase-shifted sinusoidal waves:

The figure below represents the wave of asinusoidal voltage. As we saw earlier, the mathematical formula for this wave is:

$$v = V_p \sin \theta$$

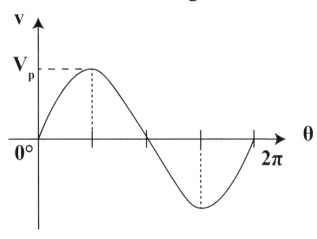

Now, when this wave is shifted to the right with respect to the reference by a certain angle <p (with a phase delay) as shown in the figure below, the formula becomes:

$$v = V_p \sin (\theta - \varphi)$$

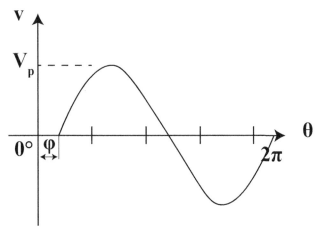

The same thing when the voltage wave is shifted to the left (figure below) by a certain angle <p (with a phase advance), the mathematical formula is given by :

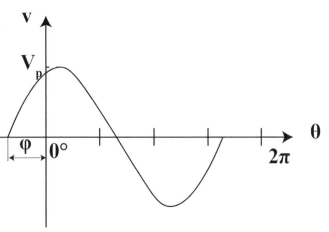

$$v = V_p \sin (\theta + \varphi)$$

III-II-VI- Instantaneous value:

From this section onwards, the alternating voltage and the alternating current are designated by the lower-case letters v and i, respectively. It is clear that v and i are functions of time as shown in the following figure, i.e., at each instant t, we have a corresponding value of current i(t).and voltage v(t) called the instantaneous value. These values are positive during the positive alternation and negative during the negative alternation.

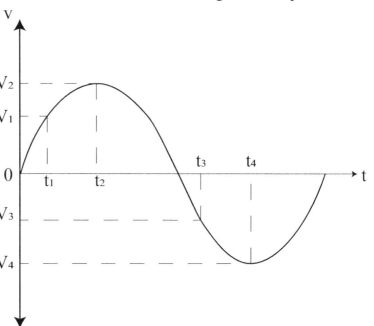

III-II-VII- Peak value:

The peak value, of a sine wave is the voltage (or current) at the maximum (amplitude), positive or negative with respect to zero, it is called V_p.

A sinusoidal signal is characterized by a single value of peak (figure opposite).

V_p is the maximum value of the voltage (current).

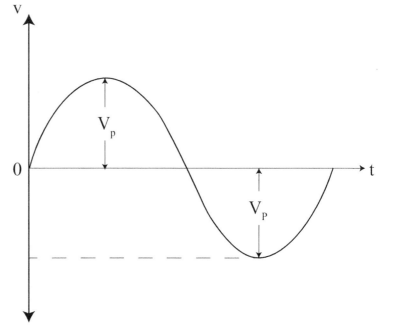

83

III-II-VIII- Peak-to-peak value:

The peak-to-peak value of a sine wave is the voltage (or current) between the positive and negative peaks. It is called V_{P-P}. It is also twice the peak value. For example, the figure opposite shows the peak-to-peak value of a sine wave voltage.

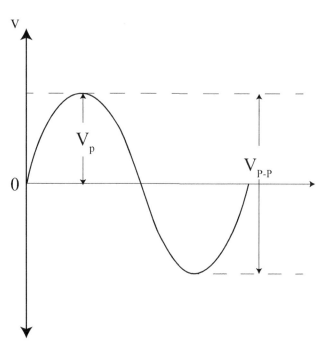

$$V_{P-P} = 2 V_P$$

III-II-IX- RMS value:

To understand what an RMS value is, you need to carefully follow the following example:

If a resistance is connected to the terminals of a battery (direct voltage source), as shown in the figure opposite. Direct voltage source. The power in this resistance gives off a certain amount of heat.

Now, we connect the same resistance to the terminals of a sinusoidal voltage source (the figure opposite), the voltage can be adjusted so that the resistance gives off the same amount of heat. Under this condition, the sinusoidal voltage has an RMS value Sinusoidal voltage source. equal to the direct voltage.

84

To be noted:

The RMS value of a sine wave voltage is equal to the direct voltage that gives off the same amount of heat in a resistance.

The peak value of a sine wave can be converted into its RMS value, for the voltage as an example:

$$V_{RMS} = V_P / (\sqrt{2})$$

Most measuring devices (voltmeters, ammeters, multimeter ...) display the RMS value. For example, 220 V in a mains socket is an RMS value.

III-II-X- Average value:

The average value of a sine wave for one period is always zero, because the values of the positive alternation cancel out those of the negative alternation. To make the average value meaningful, it is calculated over a half period rather than a full period.

Consequently:

Over one period: the average value = 0 because it is an alternative function

Over half-period : We can use the following formula to calculate the average value out of the peak value:

$$V_{AV} = 0.637 \, V_P$$

Example:

The following sinusoidal voltage would be:

Equation : $u(t) = 10\sqrt{2} \sin(315t + 1)$

Curve :

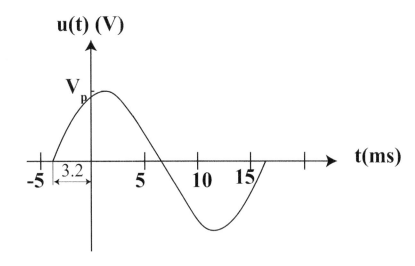

On the basis of equation we can deduce:

The pulsation: $\omega = 315$ rad.s^{-1}

The phase shift : $\varphi = 1$ rad

The period : $T = 2\pi / \omega = 20 \cdot 10^{-3}$ s

The frequency : $f = 1/T = 50$ Hz

The maximum value : $V_P = 10\sqrt{2} = 10,14$ V

The RMS value : $V_{RMS} = V_P / \sqrt{2} = 10\sqrt{2} / \sqrt{2} = 10$ V

On the basis of curve we can deduce:

One period corresponds to one turn of the trigonometric circle.

$$T \longrightarrow 2\pi$$

$$3,2 \longrightarrow \varphi$$

$$\varphi = 2\pi(3,2)/T = (2\pi(3,2)) /(20 \cdot 10^{-3}) = 1,0 \text{ rad}$$

III-III- Representation of a sinusoidal voltage:

It should be noted that the amplitude and angle of variables which vary in time, such as sine waves can be represented by a rotating vector called a phase vector (or phasor).

A complete cycle of a sine wave is represented by a phase vector that rotates 360° (or 2π rad) The figure below shows how a phase vector reproduces the sine wave ($v = V_p . \sin(\theta)$) as it rotates from O rad to 2π rad.

The positive direction of the angles

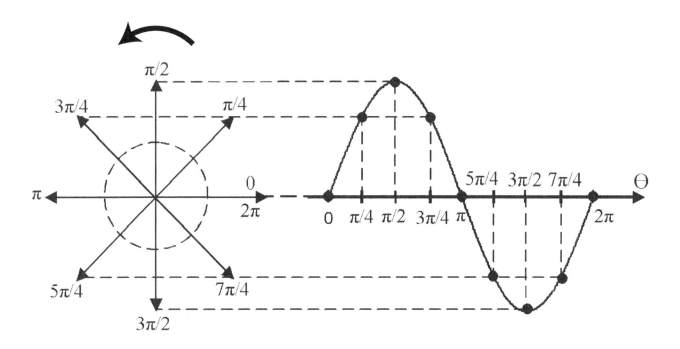

Note that the length of the phase vector is equal to the maximum value of the wave (V_p) which occurs at $\pi/2$ and $3\pi/2$.

IV- Power in sinusoidal regime :

IV-I- Instantaneous power:

Electrical power is the product of voltage by the current.

$$u(t) = U\sqrt{2} \sin(\omega.t + \varphi) \quad ; \quad i(t) = I\sqrt{2} \sin(\omega.t)$$

$$P = u\mathrm{i}$$
$$= U\sqrt{2} \sin(\omega.t + \varphi) . I\sqrt{2} \sin(\omega.t)$$
$$= 2UI \sin(\omega.t + \varphi).\sin(\omega.t)$$

To rearrange the terms, we use the trigonometric relation below:

$$\sin a. \sin b = 1/2(cos(a-b) - cos(a+b))$$

$$\Longrightarrow \quad P = U.I. \cos(\omega.t + \varphi - \omega.t) - U.I.\cos(\omega.t + \varphi + \omega.t)$$

Finally:

$$P = U.I. \cos(\varphi) - U.I.\cos(2\omega.t + \varphi)$$

We can see that the instantaneous power is the sum of a constant term "$U.I.\cos(\varphi)$" and a periodically varying term "$U.I. \cos(2\omega.t + \varphi)$".

IV-II- Active power:

The active power is the average of the instantaneous power. The average value of the periodic term is zero (it is an alternative periodic function). So only the constant term remains.

$$P = U.I. \cos(\varphi)$$

U: RMS value of the voltage (V)

I: RMS value of the current (A) ;

φ: phase shift between u and i (rad).

Unit: the watt (W).

IV-III- Reactive power:

Reactive power is a mathematical invention for facilitating calculations.

$$Q = U.I. \sin(\varphi)$$

Unit: the reactive volt-ampere VAR

IV-IV- Apparent power:

The apparent power does not take into account the phase shift between u(t) and i(t).

$$S = U.I$$

Unit: the volt-ampere V A.

IV-V- Power triangle:

By observing the above relations we can notice that:

$$S^2 = P^2 + Q^2$$

Good to know:

Only active power has a physical reality. The reactive power does not correspond to any real power.

IV-VI- Boucherot's Theorem:

- Statement of the Theorem:

The active and reactive powers absorbed by a group of dipoles are respectively equal to the sum of the active and reactive powers absorbed by each dipole.

Example :

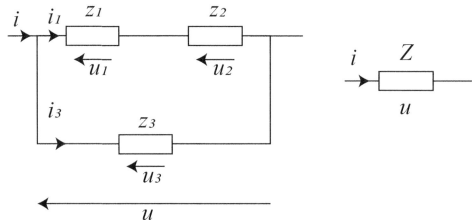

- Instantaneous power

$$P = ui = u_1 i_1 + u_2 i_2 + u_3 i_3$$

- Active power

$$P = U.I.\ cos(\varphi) = U_1.I_1\ cos(\varphi_1) + U_2.I_2\ cos(\varphi_2) + U_3.I_3\ cos(\varphi_3)$$
$$P = P_1 + P_2 + P_3$$

- Reactive power

$$Q = U.I.\ sin(\varphi) = U_1.I_1\ sin(\varphi_1) + U_2.I_2\ sin(\varphi_2) + U_3.I_3\ sin(\varphi_3)$$
$$Q = Q_1 + Q_2 + Q_3$$

Note: Boucherot's theorem is not valid for apparent power.

V- Other waveforms:

In fact, there are not only sine waves. There are other time-varying wave forms, such as triangular waves (figure A) and impulse waves (figure B). The frequency and period are determined as for the sine wave.

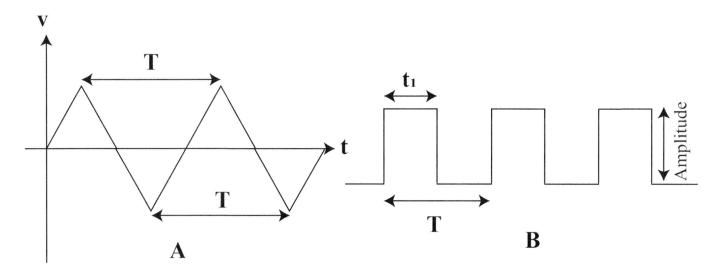

Concerning impulse waves, a very important characteristic is defined: aspect ratio (duty cycle).

To be noted:

The duty cycle is the ratio between the width of the pulse (k) and the period (T); it is commonly expressed in percent.

$$\text{Duty cycle} = (t_1/T) \times 100\%$$

Another characteristic of the pulse wave is the average value, it can be given by the following expression:

$$\text{Average value} = (\text{duty Cycle}) \times (\text{Amplitude})$$

91

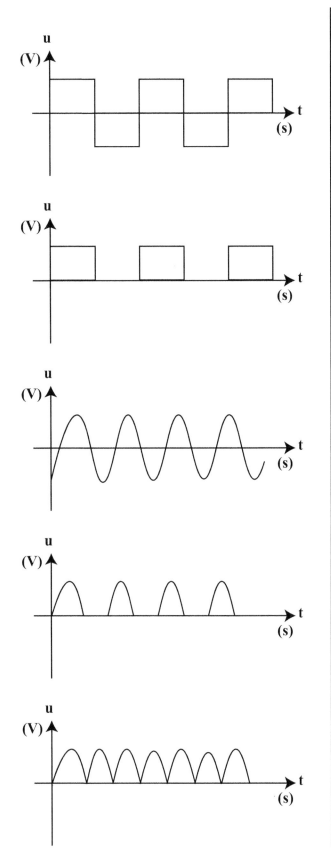

Summary table :

Signal	U_{p-p}	RMS
Symmetrical square	0	U_{MAX}
Positive square	$U_{MAX}/2$	$U_{MAX}/2$
Alternating sinusoidal	0	$U_{MAX}/\sqrt{2}$
Pulsed half wave rectifier	U_{MAX}/P	$U_{MAX}/2$
Pulsed full wave rectifier.	$2.U_{MAX}/P$	$U_{MAX}/\sqrt{2}$

Generators and their combinations.

Objectives

In this lesson we will present the different elements that we find in an electronic circuit, namely generators, resistors, capacitors, coils, diodes and transistors.

In the first lesson of this course, we will introduce a major element: the generator. It is omnipresent in any circuit or electrical installation. Firstly, we will define the generator, then we will present its different types and finally we will end with the different patterns of combining generators.

After careful study of this lesson, you will easily be able to:

- •Define what is a generator;
- •Know the types of existing generators;
- •Know the combining patterns of generators.

I. Generators:

So far, we have considered the generator as a black box providing us with voltage and current on demand. Unfortunately, generators have limitations that we will review.

To be noted:

Note that the terms source, power supply, active dipole are always associated with the term generator.

I-I- Definition:

First of all, it is necessary to know that a power supply is a device that provides a load with electrical power. A load is any element or electrical circuit connected to the output of the power supply from which it consumes current.

This figure shows a block diagram of a power supply connected to a load. This power supply provides the load with a voltage V_{out} and a current I.

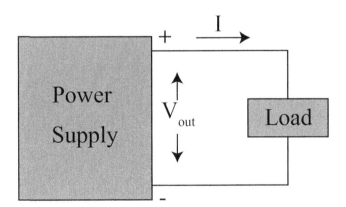

The main role of generators is to maintain the movement of electrons. They comprise two poles or terminals (positive terminal marked + and negative terminal marked -).

Among the usual generators we can mention in order of importance:

A) The power plant (involving huge powers):

NUCLEAR : The energy is provided by the fission or the explosion of a uranium or plutonium atom.

HYDRAULIC: The mechanical energy of a strong water flow is exploited: dams waterfalls, river flow.

THERMAL: Electricity is produced from the combustion of coal or oil.

B) Batteries and accumulators:

Electrical energy is obtained from chemical reactions.

C) Dynamos (direct current) and Alternators (alternating current):

They deliver an electric current from mechanical energy on the one hand and from magnetic energy on the other hand.

To be noted:

A generator, whose function is to supply energy, is called an active dipole.

A device with an input and an output (two poles) is called a dipole.

Symbols of a generator: or this one

The short thick line represents the sign.

The end of the arrow represents the terminal.

The foot of the arrow represents the terminal.

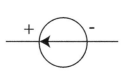

I-II- Types of generators:

There are two types of generators:

I-II-I- Voltage generators:

Definition:

Voltage generators impose a fixed voltage independently of the load. It is up to the load to determine the consumed current.

Example: Car battery, Batteries and Alternators

Symbol :

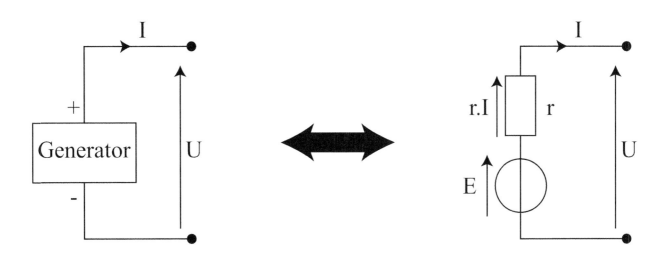

Features:

The voltage source:

1. Provides a voltage independent of the load resistance (i.e. independent of the current which the load is asked to carry).

2. Has a very low internal resistance.

General equation:

On the basis of the figure above, we can write :

$$U = E - (r \times I)$$

With :

U in V;

E: EMF = Electromotive force;

r = internal resistance in Ω;

I = current delivered by the generator in A.

Note that if r is very small, the internal voltage drop will be negligible. The voltage U in these conditions will be equal to the EMF of the source E. The generator in this case is said to be a voltage generator.

Example :

The car battery is a voltage source, it is supposed to supply always 12 V during a certain number of hours. On the other hand, it delivers a current according to the demand of the loads which it feeds.

To be noted:

Each source has an internal resistance; which will often be a factor of limitation of the performances.

a- Ideal source:

If the load can be varied to a large proportion without any variation in the voltage supplied, the source is said to be ideal. In other words, our voltage source does not have an internal resistance r = 0.

Symbol: Example:

b- Real source:

In the real world, our voltage source has an internal resistance, very low r, but present. We say that the source is real.

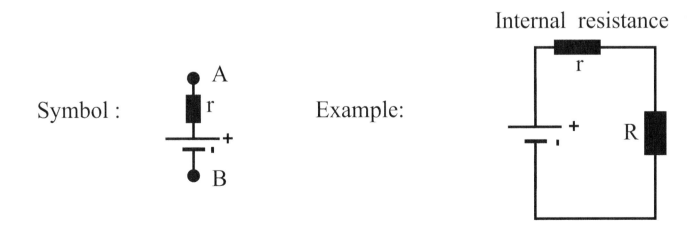

Symbol :   Example:

How can this be troublesome?

You must remember Ohm's law which says that **U = R.I.**

In the first circuit, above, the current will be identified at each point, do we agree?

This implies that the internal resistance to the source will also be crossed by the current I of the circuit and that according to this good old law, there will be a voltage drop across this internal resistance.

Still, okay?

This voltage drop will be subtracted from the voltage supplied by the source, so the load will only check U at no load (without flow) - (minus) r x I. This phenomenon is more harmful when the flow rate increases following a decrease in the value of the load, the voltage drop will increase, reducing even more the voltage available for the load.

Let's look at this with a practical example:

We have a 12 V no-load battery that has an internal resistance of 0.1 Ω. This seems reasonable, even low, wait for the next part.

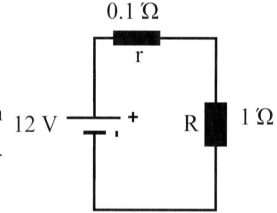

We connect on this battery a load constituted by a resistance of 1 Ω

We propose to calculate the current flowing in this circuit and the voltage at the terminals of R, which is the load.

1 - Let's calculate the total resistance:

$$R_{Tot} = R + r = 1.1\ \Omega$$

2 - Calculate I :

$$I = U / R_{Tot} = 12 / 1.1 = 10.91 \text{ A}$$

3 - Calculate the voltage UR at the terminals of the resistance R:

$$U_R = R . I$$

This gives us:

$$U_R = 1 \text{ x } 10.91 = 10.91 \text{ V}$$

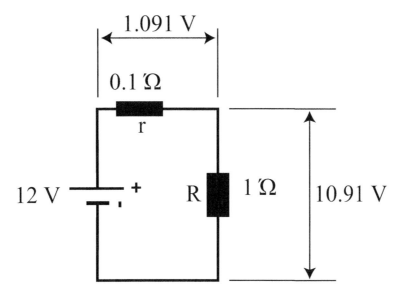

To be remembered:

1. The voltage drop at the terminals of r deprives the load (R) of this voltage.

2. The greater the internal resistance r, the greater the voltage drop.

3. The internal resistance limits the maximum flow to a value called short circuit current.

4. To obtain a maximum power transfer (U.I) this would require us to connect a load of resistance equal to the resistance of the generator.

I-II-II- Current generator:

- Definition:

Current generators deliver, under nominal conditions, a fixed current independent of the load. It is up to the load to determine the voltage.

Symbol:

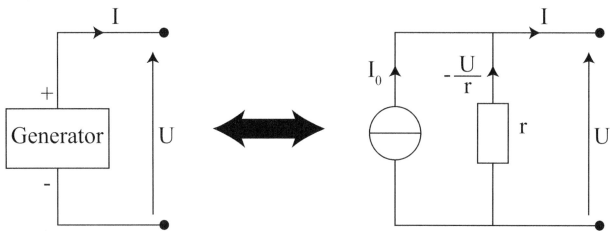

- Characteristics:

The current source :

1. Provides a current independent of the load resistance.

2. Has a very high internal resistance.

- General equation:

Let us assume that the internal resistance in current generators is always in parallel with the generator.

From the expression of the emf, we can write:

$$I = I_0 = -\frac{U}{r}$$

When r is very large, the quotient U/r tends to zero. In these conditions, the current I of the load will be equal to the current 0I delivered by the source. We thus define the current generator.

I-III- Characteristic parameters of a generator:

All generators are characterized by the following main parameters:

- An internal resistance can be measured by a multimeter position: ohmmeter;

- An electromotive force (EMF) E which can be measured, during a no-load test; by a multimeter position: voltmeter;

- A maximum current I, which can be supplied by this generator, can be measured during a short-circuit test; by a multimeter position: ammeter.

I-III-I- Electromotive force (no-load test):

When the generator does not deliver its voltage to a load (open circuit or no-load test), we can in this case define a no-load voltage for a generator, which is called the electromotive force. It is indeed an electrical force that moves (drives) the electrons. We speak of an EMF at no load E and a load voltage U.

On load, the general equation is:

$$U = E - (r\ I)$$

During the no-load test, the current is equal to zero, so the equation becomes:

$$U = E - (r\ .0) = E$$

Therefore, the voltage U in these conditions will be equal to the EMF of the source E.

I-III-II- Short circuit current (short circuit test):

In our case, let's remove the load, connect the two terminals and leave only the internal resistance r. In these conditions, we have:

$$I_{SC} = E / r$$
$$I_{SC} = 12 / 0.1 = 120 \text{ A}$$

This is the maximum current that can be supplied by this generator.

And to conclude...

Sometimes, you will face in your career a battery-powered setup that no longer works. As you are methodical, you will start immediately checking the condition of your batteries with a universal controller in the voltmeter position.

The measurement will show you that the battery is in good condition.

Now you are doubtful ... because you are about to fall victim to an internal resistance joke.

In fact, at no load, without any flow, there is no voltage drop, the voltage is intact, but as soon as you will ask for a little bit of flow to your dead battery, the voltage will drop at the terminals of the internal resistance which increases as it wears.

Morality, test on load!

Things to know:

Ampere-hour characteristic of batteries:

You know that batteries convert chemical energy into electrical energy. The ampere-hour characteristic **(Ah)** expresses the duration during which a battery provides a certain current to a load under its nominal voltage.

Example:

A 12 V battery and a capacity of 40 Ah means that the battery can deliver 40 A to a load for one hour (1 h) at its nominal voltage (12 V). This same battery will be able to deliver 8 A for five hours (5 h).

II- Combination of generators:
II-I- combination in series:

Let's imagine that we have a huge quantity of rechargeable batteries of 1.2 V that can deliver 1 A each. If our energy needs require 120 V and 45 A, we will create combinations of generators allowing us to satisfy these needs.

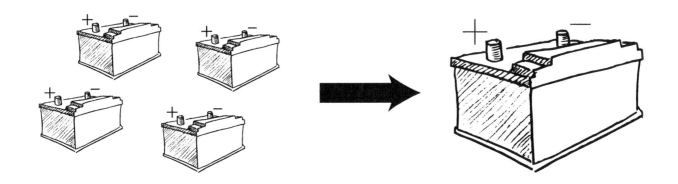

We will first define the characteristics of our generators. In all cases we will use this generator, which provides a voltage U at points A-B, which has a noted internal resistance r and which is able to deliver a current I during a time t.

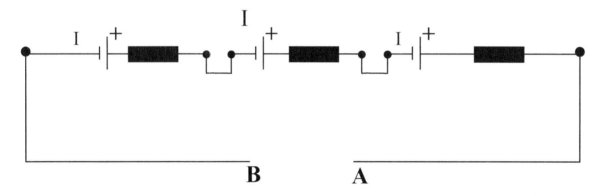

The present generator is a real one as those we meet during our tests or experiments.

Let's imagine that we put our generators in series as follows,

What are the results of such a combination?

- We discover experimentally that the voltages in this circuit add up.

- The internal resistances add up.

- The total current is equal to the maximum current that a generator can supply.

- In order to add the voltages, it is necessary to carefully connect the poles of the generators correctly, a + must be followed by a minus -.

To be remembered:

When generators are connected in series :

We wish to know the voltage available at no load at the terminals of a combination of 3 batteries in series, the maximum flow rate allowed by this assembly and the total internal resistance.

Solution:

We know that the voltages add up so it results for 3 batteries:

$$U_{tot} = 1.5 + 1.5 + 1.5 = 4.5 \text{ V}$$

$$I_{max} = I \text{ max of a battery or } 150 \text{ mA}$$

$$r_{tot} = 0.01 + 0.01 + 0.01 = 0.003 \text{ } \Omega$$

II-II- Parallel combinations (//):

What happens if we now connect our generators as shown in the diagram ?

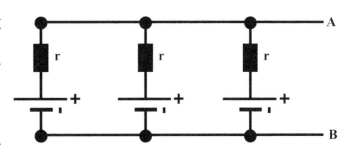

- We notice that the internal resistances are in parallel, which decreases the global internal resistance.

- The total voltage (assuming the generators are identical) equals the voltage of one generator.

- The total current available is equal to the sum of the currents that can be supplied by the generators.

To be remembered:

In the settings // of generators:

I total = n times the current of a generator ;

U total = U of a generator ;

r total internal = divided by n times the number of generators.

What are the advantages and disadvantages of such a combination?

- Advantages:

- The available current is multiplied by a factor n = number of generators;

- The internal resistance of the assembly is divided by a factor N equal to the number of generators;

- A H.S. generator does not disturb the assembly very much.

- Disadvantages:

- The overall voltage is dictated by the unitary voltage of the generators.

II-III- Combination in opposition:

There is nothing to prevent us from making a connection of generators such as the one you can see on the left. It can be a mistake or on the contrary a deliberate action. BUT...

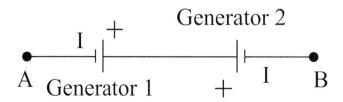

In the above case, the voltages are in opposition and will cut each other off , this can lead us to 3 simple cases:

1. Generator 1 provides a higher voltage than generator 2, in such a case

it will impose the direction of the current flow in the circuit. The resultant will be $U_1 - U_2$.

2. The generator 1 supplies a voltage equal to the generator 2, the resultant is null, no current circulates in the circuit.

3. Generator 1 supplies a lower voltage than generator 2, in such a case the generator B will impose the direction of current flow in the circuit, the resultant will be $U_2 - U_1$.

Replacing or adding a battery:

Replacing one battery with another does not cause any particular problem. On the other hand, if you want to add one or more batteries to the engine, it is necessary to :

1- Take into account:
- The characteristics of the batteries set (equivalent output voltage and usable capacity);
- The location of the terminals.

2. Follow these few rules:
- Never swap the positive and negative terminals when assembling;
- Always isolate the batteries when working on the engine;
- Never install two batteries of different voltages in parallel;
- Never install in series two batteries with different capacities.

Reminder:

Wiring is said to be in series when the negative pole of one battery is connected to the positive pole of the next battery.

In this type of wiring:

- The output voltage is equal to the sum of the voltages of the batteries.

- The capacity remains that of a single battery.

Connect the (-) pole

with the (+) pole

24 V - 100 Ah

Two 12 V - 100 Ah batteries connected in series give us a:

- Double voltage: 12 V + 12 V = 24 V

- Capacity equals 100 Ah

The wiring of the batteries is said to be in parallel when all the minus and plus poles of the different batteries are connected together.

In this type of wiring:

- The output voltage remains constant and equivalent to the voltage of a single battery.

110

- The usable capacity is the sum of the capacities of the different batteries.

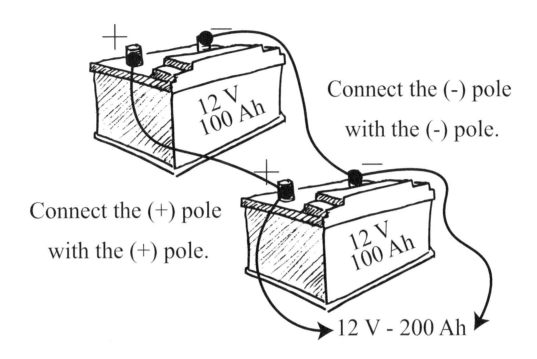

These same batteries mounted in parallel, give us a :

    - Double capacity: 100 Ah + 100 Ah = 200 Ah

    - Equal voltage: 12 V

As for the wiring of the batteries in series and in parallel as shown in the figure below, we obtain:

- Increase of the output voltage compared to the voltage of the other batteries;

- Increase of the usable capacity compared to the capacity of the different batteries.

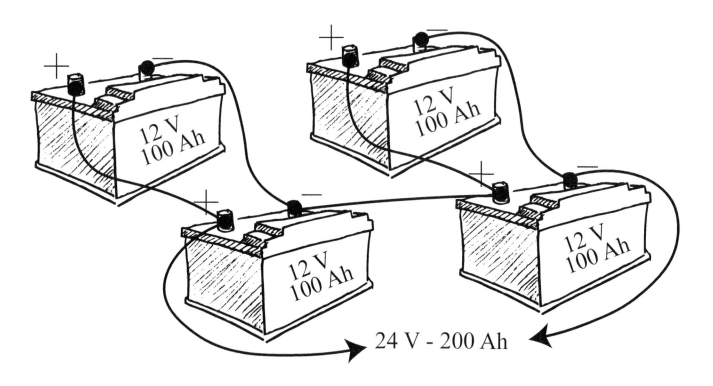

24 V - 200 Ah

The combination of the series and parallel groupings gives us a :

- Increase in voltage: 12 V + 12 V = 24 V

- Increase in capacity: 100 Ah + 100 Ah = 200 Ah

Resistance.

Objectives:

In this lesson, we will present the different elements found in an electronic circuit, namely generators, resistors, capacitors, coils, diodes and transistors, we will also describe resistors: definition, characteristics, types and combinations in series, parallel or series-parallel.

After careful study of this lesson, you will easily be able to:

- Know the resistances in an electrical circuit
- Calculate the equivalence of a series combination;
- Calculate the equivalence of a parallel combination;
- Calculate the equivalence of a series-parallel combination;
- Know the different uses of resistors.

I- Definition:

In this lesson, we study the resistance of a homogeneous conductor, i.e. the opposition that a material exerts to the passage of current I. A homogeneous conductor is a body that has the same properties throughout its volume. Resistance is of paramount importance in electricity and can be said to be a constant in the study of electrical circuits.

Explanation:

The movement of electrons from an atom requires a force capable of overcoming the attractive force of the nucleus. When the electron moves in matter, it causes collisions with other atoms, which decreases the amount of the moving electrons. This opposition to the movement of electrons causes electrical resistance. Therefore, the property of a material to oppose the flow of current is called **resistance**.

Terms used: Resistances;

Notation: R;

Unit: Ohm;

Symbol: Ω (omega);

Graphic symbol: ─▭─ Or: ─/\/\/\─

The instrument used to measure electrical resistance: Multimeter position: Ohmmeter.

II- Types of resistors:

There is a wide variety of resistors on the market. They differ in shape and size. Despite this, they can generally be classified into one of two categories: fixed or variable.

II-I- Fixed resistors:

They have factory-determined values (unmodifiable).

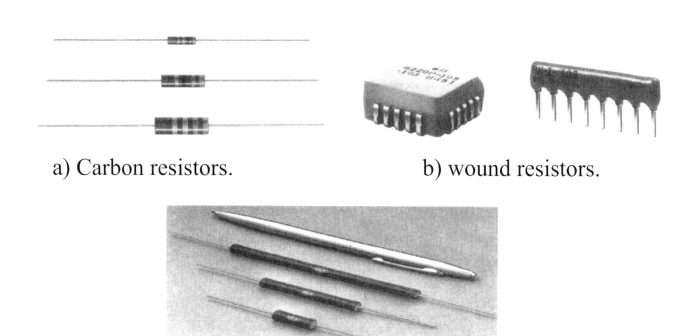

a) Carbon resistors.   b) wound resistors.

c) Resistor networks.

How are fixed resistors made?

Fixed resistors are made using various methods and materials. Typical examples are shown in the previous figure.

Generally, there are two types of fixed resistors, wound resistors (high power) and film or carbon resistors (low power).

A. Wound resistors:

They consist of a metal wire (often a nickel-chromium alloy) wound on a porcelain or soapstone* support.

* Compact silicate rock.

Their resistance is given by the relation that you now know:

$$R = \rho \cdot \frac{l}{S}$$

B. film resistors:

They consist of a cartridge track deposited by pyrolysis on a support, the whole setup is covered with a varnish and a synthetic resin.

The section and length of the track determine the value of the resistance.

Marking of the fixed resistors:

The ohm value of the resistors is indicated either; in plain text on the body or by the color code.

In fact, for wound resistors, the value and the maximum power are generally read directly on them. For carbon resistors, the value of the resistance is determined by means of a color code.

A color-coded resistor consists of colored bars. Each color is assigned a value from 0 to 9.

Each color corresponds to a number; only the first 3 colors are significant in determining the resistance value. The last ring corresponds to the tolerance, it is either further away from the other rings or thinner.

The reading must be done starting with the first ring, which is the closest one to an end.

Resistor

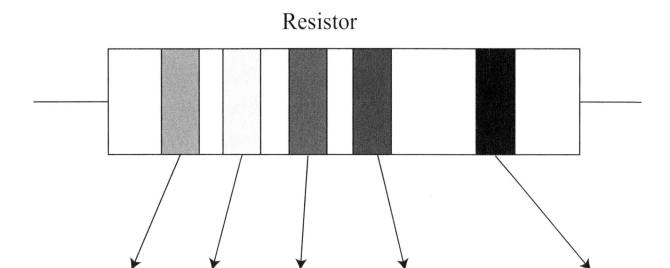

Color	1st Digit	2nd Digit	3rd Digit	Number of zero.	Tolerance in %.
Black	-	0	0	-	-
Brown	1	1	1	0	± 1%
Red	2	2	2	00	± 2%
Orange	3	3	3	000	-
Yellow	4	4	4	0 000	-
Green	5	5	5	00 000	± 0.5%
Blue	6	6	6	000 000	-
Purple	7	7	7	-	-
Gray	8	8	8	-	-
White	9	9	9	-	-
Gold	-	0	0	x 0.1	± 5%
Silver	-	0	0	x 0.01	± 10%

To be noted:

Without color the tolerance is ± 20 %.

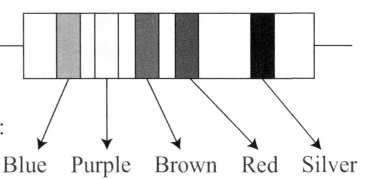

An example to understand:

Here is a resistor, Figure opposite:

Blue   Purple   Brown   Red   Silver

The ring on the right, which we want to represent in silver color represents the tolerance of the value, i.e the percentage deviation around the value indicated. Silver = 10%.

Let's start with the first ring, it is in blue, (blue = 6), the next is purple (purple = 7) and the third, which is the multiplier, is in brown (brown = 1), red (number of zero) = 00.

Now, here are some remarks:

Remark 1- The values of the resistors are normalized:

We do not find in the trade any value; these belong to a series based on the nth root of 10. Here below the series E12:

10 - 12 - 15 - 18 - 22 - 27 - 33 - 39 - 47 - 56 - 68 - 82 - 100

We naturally find in this series all the multiples of 10, which would give:

100-120-150-180-220-270 ... Etc.

Note 2- Resistors have a value tolerance:

If you read 4700 on the body of the resistor, don't be surprised. If you measure it with a trustworthy instrument, you will not find that value because the manufacturers display a manufacturing tolerance.

The resistors we use in our field have a tolerance between 5 and 10%. We can use values of 1% but it will be more expensive...

Often, you will find the resistance value and its tolerance stamped in plain text on the body of the element. Thus, in a certain system,

R stands for decimal point and the letters correspond to the following tolerances:

$$F = \pm 1\% \qquad G = \pm 2\% \qquad J = \pm 5\% \quad K = \pm 10\% \qquad M = \pm 20\%$$

Also note that for ohmic values less than 100 Ω, R serves as a decimal point designator.

For example:

- **6R8M**: Means a resistance of 6.8 Ω with an accuracy of $\pm 20\%$.
- **3301F**: Means a resistance of 3300 Ω with an accuracy of $\pm 1\%$.
- **2202J**: Means a resistance of 22000 Ω with an accuracy of $\pm 5\%$.

Note: The resistors support a maximum power:

Beyond which ... they burn out. As their function is to slow down the passage of current, they convert this energy into heat. We must therefore use a resistor dimensioned for the mission we entrust to it, that is to say, to ensure its maximum power. The usual values in electronics are 1/8 W, 1/4 W, 1/2 W.

II-II- Variable resistors:

A- Linear variable resistors:

These are resistors whose value can be modified by the user according to the needs. The value follows Ohm's law.

They are used in two fundamental applications:

- Current control (rheostat), the graphic symbol is: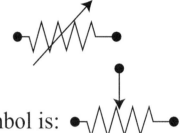

- Voltage division (potentiometer), the graphic symbol is: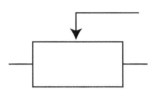

A-I- Rheostat:

This is an adjustable resistor with only 2 terminals; we essentially change the intensity.

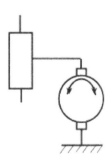

Example: on the figure opposite, the fan engine is placed in series with the rheostat; if R is low, the intensity will be high and the engine will turn quickly, if R is high, the intensity will be low and the engine will turn slowly.

A-II- Potentiometer:

Potentiometers are variable resistors (with a cursor) that allows to adjust a voltage to the desired value.

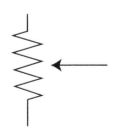

Principle:

This three-terminal device allows the resistor to be continuously varied from maximum to minimum value. A potentiometer is used to control the output voltage (Vo), because if a fixed voltage (U) is applied to the two ends of a resistor, a variable output voltage (Vo) will be drawn.

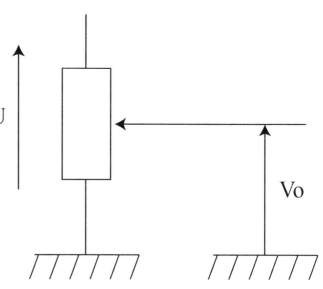

This voltage is proportional to the value of the resistance between the cursor and the mass.

In this assembly, the voltage is modified, hence the name of voltage divider bridge.

Cursor position	r	Voltage Vo
On top	= R	12 V
In the middle	= 0.5 R	6 V
In the bottom	= 0	0 V

Applications:

The most common uses are mounted as:

- Every day you act (or used to act) on a resistor without knowing it by fiddling with the volume knob on your TV, HI FI system or radio.

- Accelerator pedal sensor; the manufacturer has memorized several voltage points in the ECU; the time between these points allows the ECU to know how fast the accelerator pedal is depressed in order to correct the suspension to prevent the vehicle from carburizing.

B- Non-linear variable resistors:

These are metallic or semi-conductive materials whose resistance varies either with temperature (thermistor) or with light (photoresistor) but in non-linear ways.

B-1- Thermistors:

These are thermosensitive resistors which are of 2 types, NTC and PTC.

Thermistors with negative temperature coefficient (NTC):

It is a semi-conductor whose resistance decreases rapidly when the temperature increases

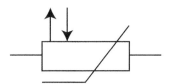

When the temperature increases, (T ↑),

the resistance decreases (R↓).

Thermistors with positive temperature coefficient (PTC)

A PTC thermistor is a semi-conductor whose resistance increases rapidly when the temperature increases.

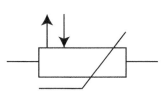

As the temperature increases, (T ↑),

the resistance increases (R ↑).

If we translate this on a graph, we obtain:

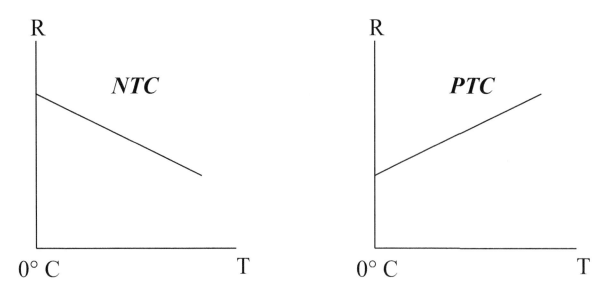

Applications:

NTC thermistors are primarily used in automotive applications as a temperature sensor?

- Of the air between the evaporator fins of air conditioners;

- Of the intake air in fuel injection systems;

- Of water in the cooling system (to adjust the injection time);

- Of oil and/or water to inform the instrument panel indicator.

B-2- Photoresistors:

It is a semi-conductor whose conductivity varies according to the amount of light it receives; photoresistors are used like detection or measurement of illumination.

III- Combinations of resistors:

III-1- Resistors in series:

a - Formula for two resistors:

When two or more resistors are mounted next to each other, they are said to be in series. A series circuit provides a single path for the current flow.

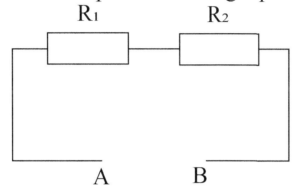

Manipulation:

We drew a resistor of $R_1 = 10\ k\Omega$ and another of $R_2 = 4.7\ k\Omega$ from our components stock. We have connected them as shown in the diagram above. We are supposed to determine the resulting value (total).

As we know, a resistor slows down the current flow, so if we put these two resistors end to end, (this is called, combination in series) we will increase the value of the brake. Translated into mathematical language, the equivalent resistance is the sum of the two resistors connected in series:

$$\mathbf{R_T = R_1 + R_2}$$

In this case we will have: $R_T = 10 + 4.7 = 14.7\ k\Omega$

To be noted:

Note that we do not add kilograms and tons, we usually use a common unit, in this case $k\Omega$.

What if we had 5 resistances connected at the ends of each other?

No problem, the brakes add up, the resultant would be:

$$R_T = R_1 + R_2 + R_3 + R_4 + R_5.$$

There is no difficulty with the series grouping of resistors.

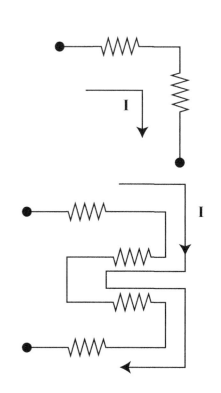

To be remembered:

If there is only one path for the current between two points (figure opposite) of an electric circuit, the resistors between these points are in series, whatever the way they are presented; and the equivalent resistance of such a combination is the sum of the resistances.

Example:

In the circuit shown in the figure opposite we find a generator, in this case a 10 V battery, a 3 kΩ resistor and another of 7 kΩ. The values, unconventional, were chosen to simplify the calculations.

We want to calculate the total (equivalent) value of the resistances in series:

Rt = R1 + R2

Rt = 3000 + 7000 = 10 000 Ω or 10 kΩ

The next Figure represents the equivalent circuit of the previous circuit. We can easily calculate the current flowing in the circuit as well as the voltage drops at the terminals of each resistor.

Starting with the current flowing in this circuit. By applying Ohm's law, $I = U / R$, we obtain:

$I = 10/10{,}000 = 0.001\ A = 1.\ 10\text{-}3\ A = 1\ mA$.

This is where it gets interesting, we know that the current flowing is identical at every point in this circuit, i.e. the current flowing through the 3 kΩ resistor is identical to the one flowing through the 7 kΩ resistor.

Now, let's calculate the voltage drops at the terminals of resistors R_1 and R_2:

At the terminals of the resistors combination, there is a 10 V generator, but it is not the same at the terminals of each resistor. We know that $U = R\ I$, let's apply this.

- On the 3 kΩ resistor:

$$U_{R1} = R_1\ I = 3000 \times 0.001 = 3\ V$$

- On the 7 kΩ resistor:

$$U_{R2} = R_2\ I = 7000 \times 0.001 = 7\ V$$

b) General formula

When a circuit has more than one resistor of the same value in series, there is a quick method to get the total equivalent resistance. Simply multiply the ohmic value of a resistor by the number of resistors connected in series

series with the same value.

$$R_{TOT} = n \times R$$

Where:

n is the number of resistors of equal;

R is the ohmic value of a resistor.

III-II- Resistors in parallel or derivation (//):

a- Formula for two resistors:

When two or more resistors are connected between the same two points, they are said to be in parallel with each other. A parallel circuit offers more than one path for the current flow.

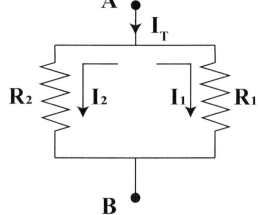

Manipulation:

We drew a 10 kΩ resistor and a 20 kΩ resistor from our components stock.

We connected them as shown in next figure.

We are supposed to determine the resulting value.

This time you will be surprised, because putting two resistors in parallel makes the total resistance decrease.

How to explain such a mystery?

Put yourself in the place of the current for a moment. You see these two resistors in front of you (point A), you are going to rush into the one with

128

the least resistance, which will determine a current I_1 but you will still have to cross the one with the greatest resistance, which will create a current I_2. The courses result: The total current will be equal to I_1+I_2, or the total current will be higher than the highest current of a branch in any case, which hopefully shows that the value of 2 resistors in // is lower than the lowest of the two values!

Let's see this with a mathematical formula:

$$R_T = \frac{R_1 \times R_2}{R_1 + R_2}$$

To be noted:

This formula is valid only for the case of two resistors connected in parallel.

Academic approach:

To calculate the resulting value of a resistors combination R1 and R2 put in //, we will go through a calculation using the inverse of the resistance which is called the conductance.

Reminder:

Remember, reducing to the same denominator is obligatory to calculate the values of n resistors in //.

As a comparison, if you are given 1/8 of a pie and 1/2 of a pie, it is difficult to see what this represents.

Mentally you will convert 1/2 into x/8 (actually it is 4/8), then you will add 1/8 which will give you 5/8 pie.

$$\frac{1}{R_T} = \frac{1}{R_1} + \frac{1}{R_2}$$

(Now you have to reduce to the same denominator)

1x

$$\frac{1}{R_T} = \frac{1 \times R_2 + 1 \times R_1}{R_1 \times R_2} = \frac{R_2 + R_1}{R_1 \times R_2}$$

We have $1/R_T$ and we want R_T so let's take the inverse, it gives:

$$\boxed{R_T = \frac{R_1 \times R_2}{R_1 + R_2}}$$

Example:

Suppose we have 3 resistors in // whose values are:

$R1 = 50 \ \Omega$, $\quad$ $R2 = 100 \ \Omega$ $\quad$ and $\quad$ $R3 = 200 \ \Omega$.

What is the equivalent resistance? (We already know that the value will be $< 50 \ \Omega$)

Let's reduce to the same denominator

$$\frac{1}{R_T} = \frac{1}{50} + \frac{1}{100} + \frac{1}{200}$$

Let's reduce to the same denominator:

$$\frac{1}{R_T} = \frac{1 \times 50 \times 100}{50 \times 100 \times 200} + \frac{1 \times 50 \times 100}{100 \times 200 \times 50} + \frac{1 \times 50 \times 100}{200 \times 100 \times 50}$$

$$\frac{1}{R_T} = \frac{35}{1000}$$

Now we know that $1/R_T = 35/1000$, let's reverse the ratio to get R_T:

$$R_T = 1000/35 = 28.57 \; \Omega$$

To be remembered:

When resistors are connected in parallel, the total resistance of the circuit decreases.

The total resistance of a parallel circuit is always smaller than the smallest one among all resistors. For example, if a 10 Ω resistor is connected in parallel with a 100 Ω resistor, the total resistance will be less than 10 Ω.

b- General formula:

When a number n of resistors with the same value are connected in parallel, the total resistance can be quickly calculated using the following formula:

$$R_T = R / n$$

This is not everything...

What about the voltage and the current in this combination?

For the voltage, we can clearly see that it is identical on each branch

(Figure opposite), ie the voltage at the terminals of R_2 equals the voltage at the terminals of R_1. This will be a constant on each branch.

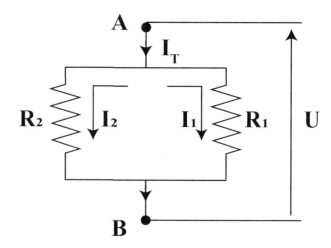

Regarding the current in each resistor, we always apply Ohm's law:

$$I_1 = U / R_1 \quad ; \quad I_2 = U / R_2$$

And to conclude, a practical example: There is a generator (battery) of 60 V feeding a combination of two resistors in parallel.

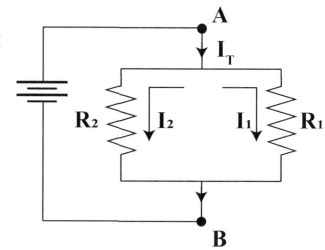

$$R_1 = 100 \ \Omega$$
$$R_1 = 150 \ \Omega$$

What are the currents flowing through each resistor and what is the current supplied by the generator?

We can either calculate the current in each branch and sum it up, or calculate the equivalent resistance, then the total current, then the current in each branch. Let's adopt solution 1.

$$I_1 = U / R_1 = 60 / 100 = 0,6 \ A$$
$$I_2 = U / R_2 = 60 / 150 = 0,4 \ A$$

The total current is: It = 0.6 + 0.4 = 1 A

Let's check our calculations:

$$R_T = \frac{R_1 \times R_2}{R_1 + R_2} = \frac{100 \times 150}{100 + 150} = 60 \ \Omega$$

$$I_T = \frac{U}{R_1} = \frac{60}{60} = 1 \ A$$

To Know:

So, you should always bear in mind that each current path in a parallel circuit is called a branch; and the voltage at the terminals of each branch is equal to the voltage at the terminals of the other branches in parallel.

That's it.

This concept of parallel resistors is simple, master it completely.

III-3- Resistors in series-parallel:

We know how to calculate the equivalent resistance of n resistors in series, we are able to do the same thing for resistors in //, we will apply our knowledge on a series-parallel combination.

Let's determine the equivalent resistance at points A and B of this assembly (the next figure):

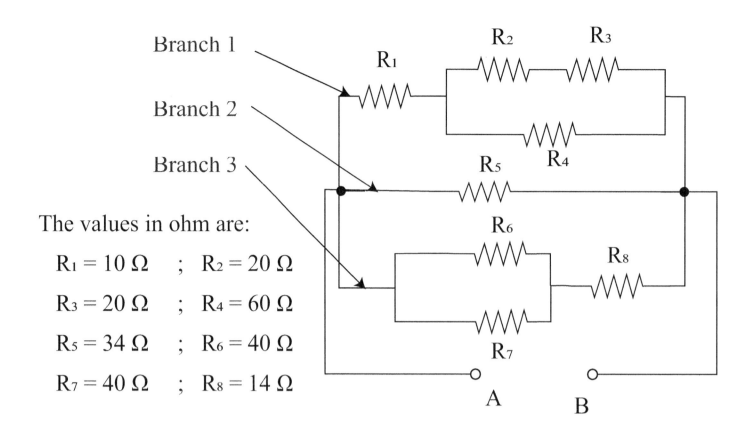

The values in ohm are:

$R_1 = 10 \ \Omega$; $R_2 = 20 \ \Omega$

$R_3 = 20 \ \Omega$; $R_4 = 60 \ \Omega$

$R_5 = 34 \ \Omega$; $R_6 = 40 \ \Omega$

$R_7 = 40 \ \Omega$; $R_8 = 14 \ \Omega$

How to proceed?

It is easy to see in this example that we are dealing with a combination that comprises three main branches in parallel:

- The one formed by R_1 R_2 R_3 R_4;

- The one formed by R_5;

- The one formed by R_6 R_7 R_8.

Within each branch, there are serial-parallel combinations that are easy to calculate.

Let's calculate the branch n°1: We have R_2 and R_3 in series R_2-R_3 is in // with R_4 This setup is in series with R_1.

$$R_2 + R_3 = 20 + 20 = 40 \ \Omega$$

$$(R_2 + R_3) \ // \ R_4 = 24 \ \Omega$$

$$((R_2 + R_3) // R_4) + R_1 = 34 \ \Omega$$

Let's calculate the branch n°2:

1 single resistor R_5 :

$$R_5 = 34 \ \Omega$$

Let's calculate the branch n°3:

R_6 is in // with R_7;

$R_6//R_7$ in series with R_8.

$$R_6 // R_7 = 20 \ \Omega$$

$$(R_6 // R_7) + R_8 = 34 \ \Omega$$

Here is our assembly reduced to 3 branches of 34 Ω in parallel which gives: 11,3 Ω

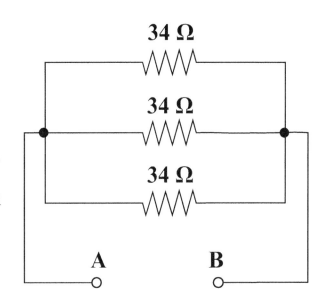

The resistors in electrical or electronic assemblies (and they are numerous) are used either to limit the current flowing in a circuit, or to divide the voltage or to release heat.

IV- Behavior of a resistor in an electronic circuit:

Usually, in a direct or alternating current electronic circuit, the resistor does not cause any phase shift between the current flowing through it and the voltage at its terminals.

Example:

The phase shift between the voltage and the alternating current caused by the resistor is zero. We say that the current and the voltage are in phase, (They reach their maximum and their minimum at the same time).

Introduction to electrical circuits.

Objectives:

This lesson is an introduction to electrical circuits. Therefore, it is important to know firstly what an electric circuit is; and to define all its components.

After a careful study of this lesson, you will be able to easily:

- Define what an electrical circuit is;

- Know the different types and layouts of an electrical circuit;

- Define a ground circuit;

- Know how to protect an electrical circuit.

I- Nature of electrical circuits:

I-I- Basic electrical circuit:

An electrical circuit in its simplest form must consist of a device producing or supplying current (example: a battery), a load, or receiving (example: a lamp), and conductors (electrical wires) that carry the electric current. The connections between these elements, constituting an electrical circuit, are called electrical terminals.

Electrical symbols:

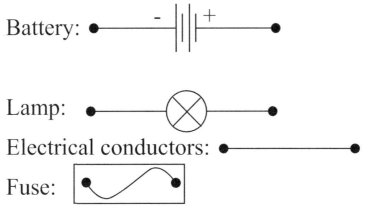

Battery:

Lamp:

Electrical conductors:

Fuse:

Switch:

In practical situations, it is essential to be able to cut off the electric current in a circuit. For this reason, we always add a control device, a switch, to interrupt or restore the flow of current in the electrical circuit. A switch has two positions, it can be either closed or open

Electrical symbol:

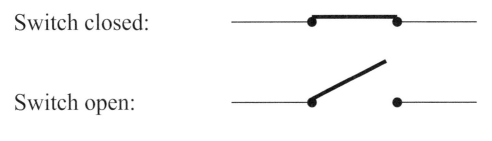

Switch closed:

Switch open:

Fuse: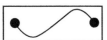

The fuse is a protection component of the electric circuit elements. In fact, a fuse is incorporated into the circuit to prevent any damage in consumer devices and to avoid burning the electrical conductors when the amount of current increases suddenly in the electrical circuit.

It is then the fuse that melts to provoke an automatic opening of the electrical circuit.

For example, the figure below represents an elementary electrical circuit that includes a battery (the electrical source), a switch, a fuse and a lamp (the load). The connection between these two elements is ensured by conducting wires.

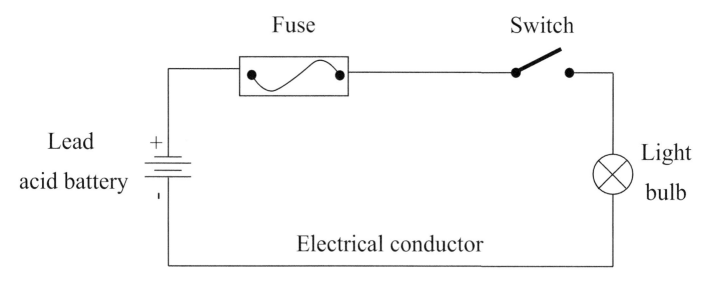

Basic electrical conductor

I-II- Ground circuit:

Taking the example of the previous figure, it is clear that the current leaves the positive terminal towards the lamp; this represents the forward path.

Then, the current leaves the lamp and enters into the battery through the negative terminal; this represents the return path. Usually this is called the circuit ground. In many practical cases, the negative terminal of the battery is connected to a grounding point. For example, in an automobile, the negative terminal of the battery is connected to the metal chassis of the vehicle.

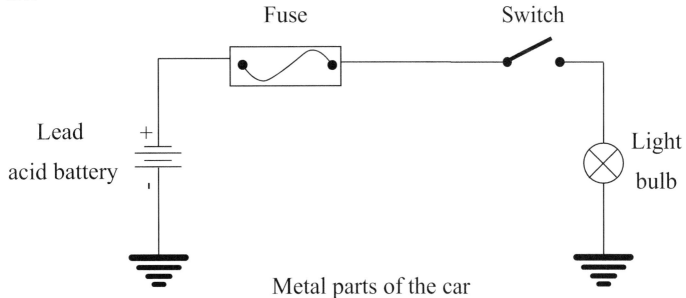

Electrical circuit with ground return

II- Types of electrical circuits:

Electrical circuits are divided into two major types:

    1. Electronic circuits

    2. Electrical circuits.

II-I- Electrical circuits:

Electrical circuits are characterized by their function which is strictly linear resistance.

In other words, a circuit is electrical as long as its function is to transfer energy.

There are two main types of circuits, the second type is subdivided into two groups

    - The basic circuit;

    - The controlled circuit.

        * The electrically controlled circuit;

        * The electronically controlled circuit.

Basic circuit:

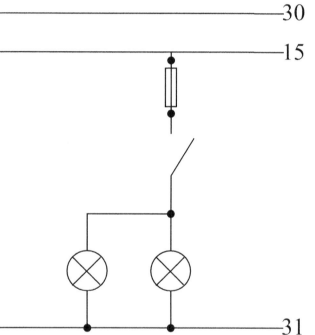

A basic electrical circuit is a circuit that uses a voltage source, a fuse, a switch, a consumer, and a ground, i.e., the switch directly controls the consumer electrically. All the current consumed by the circuit passes through the switch.

Example: brake switch, reverse lights.

Controlled circuit:

Electrically controlled circuit:

An electrically controlled circuit comprises two electrically separate circuits:

- A control circuit that uses a voltage source, a fuse, a switch, small conductors, a relay winding and a ground;

- A load or power circuit that uses a voltage source, a fuse, the mechanical contacts of the relay, larger section conductors, a consumer and a ground.

Example. horn circuit:

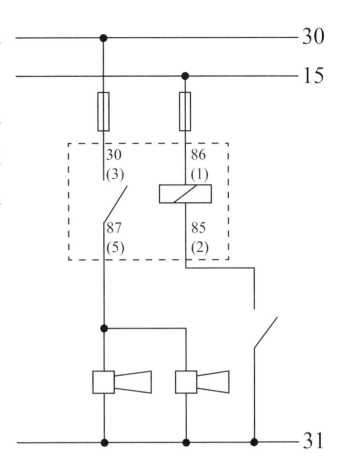

II-II- Electronic circuits:

Electronic circuits are distinguished from electrical circuits by the control they exert over electrical energy. The main electronic circuits are electronically controlled and receive impulses resulted from an electronic control module; transistor circuits and integrated circuits.

The control of such a circuit requires special knowledge and a lot of precautions to avoid damaging the circuits.

Example: flashing box, alarm system, ...

III- States of electrical circuits:

III-I- Open circuit:

An electric circuit is said to be open when its switch is open. Consequently, in an open circuit the current no longer flows. The current flow can be interrupted voluntarily by a switch or accidentally by unspecified defect. In both cases, the circuit is said to be open.

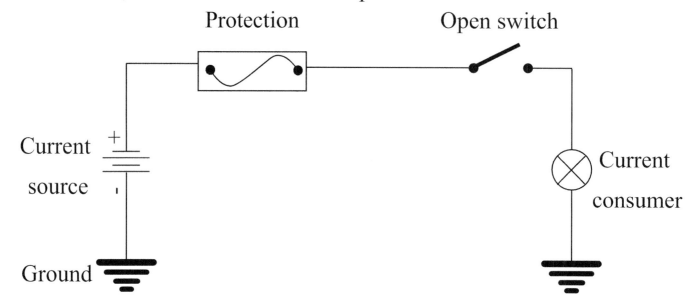

Voluntarily open circuit

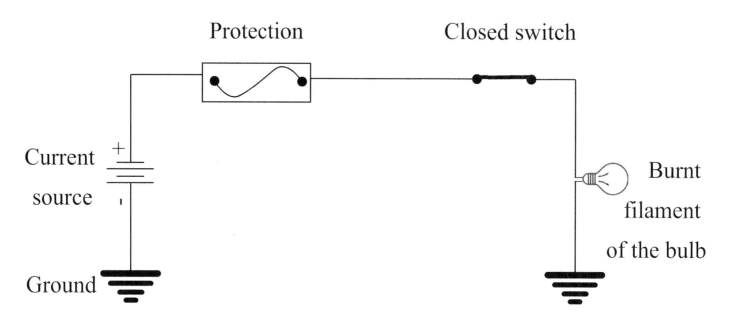

Electrical circuit accidentally opened

III-II- Closed circuit:

If the circuit is powered and its switch is closed now, the current flows through all the elements of the electrical circuit so as to return to the source, the circuit is said to be closed. There is consequently a current flow. We simply say that the circuit is energized.

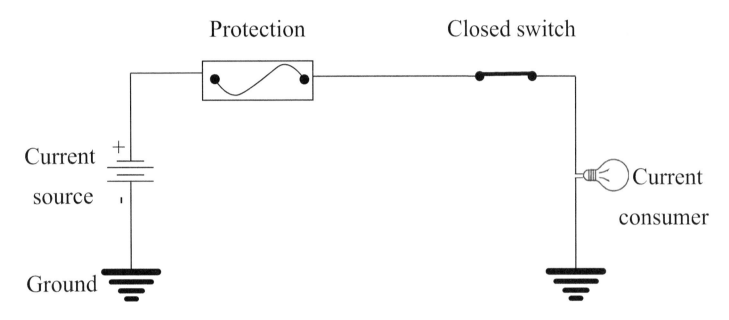

Closed electrical circuit

III-III- Short circuit:

A short circuit occurs when a bare wire touch ground due to poor isolation. In such a case, the electric current can take a path of very low resistance in order to return to the source, bypassing the load as well (electrical device). Under these conditions and without using a fuse, it is certain that the envelope of the conductor would burn due to an over increase in current.

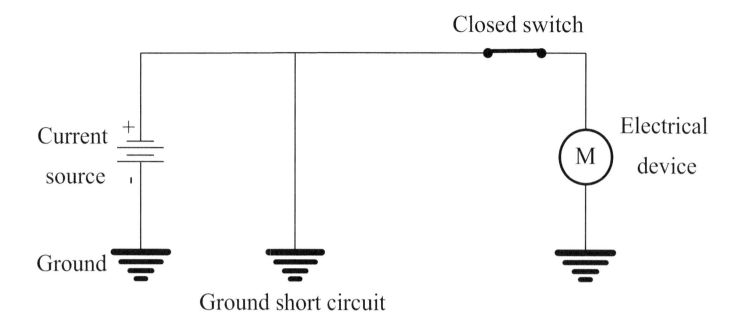

Ground short circuit without fuse

In order to protect the elements of the circuit, a fuse is incorporated. It is thus the fuse that melts to provoke an automatic opening of the electrical circuit.

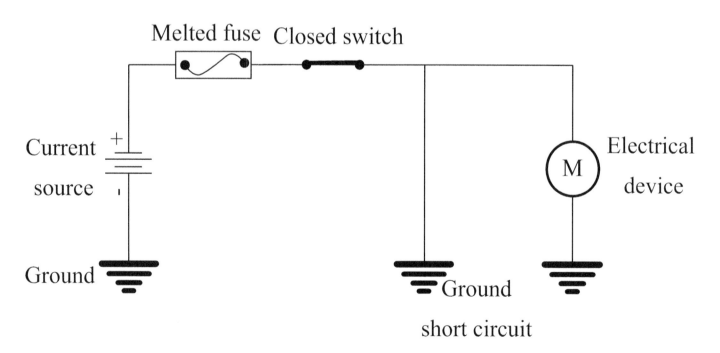

Ground short circuit with fuse

Example:

For example; using the metal ground of the automobile can present a certain danger. If the power supply conductor of an operating device accidentally touches the ground, the current is short-circuited. The current then flows directly back to the source through the contact point X, which is considered by definition as a short circuit.

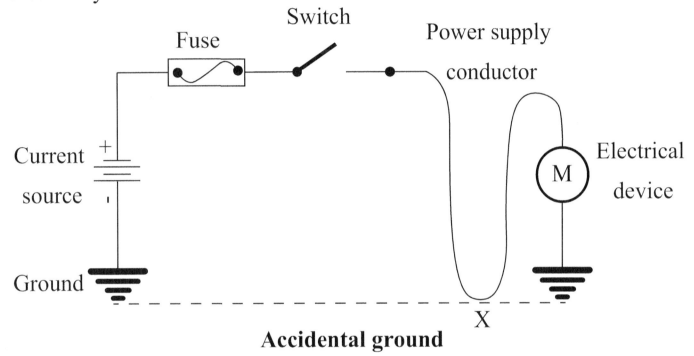

Accidental ground

IV- Layout of electrical circuits:

An electrical circuit is a continuous path formed by a conductor that runs from the source of electrical supply current, to the protection, then to the switch and finally to the load and then returns to the source. Electrical circuits can be arranged in three ways: in series; in parallel; or in parallel series.

IV-I- Series circuit:

In a series circuit, all the electrical consumers are arranged one after the other. The current flows through each consumer and returns to the source via the ground.

Resistance in a series circuit:

In such an arrangement, each consumer offers resistance to the flow of current. Therefore, the more consumers there are in series, the higher the total resistance of the circuit.

Current in a series circuit:

Since there is only one possible path for current in a series circuit, the current is the same for all consumers. If an interruption occurs, the current will not be able to flow and all consumers in the circuit will be inoperative.

Voltage in a series circuit:

The voltage in a series circuit can be increased or decreased by varying the resistance in the circuit. The higher the total resistance of the circuit, the lower the voltage due to the voltage drop at each consumer and vice versa.

This principle is used on automobiles to control the current at the consumers.

Example:

The intensity of the dashboard lighting is controlled by increasing or decreasing the resistance in the circuit by means of a rheostat.

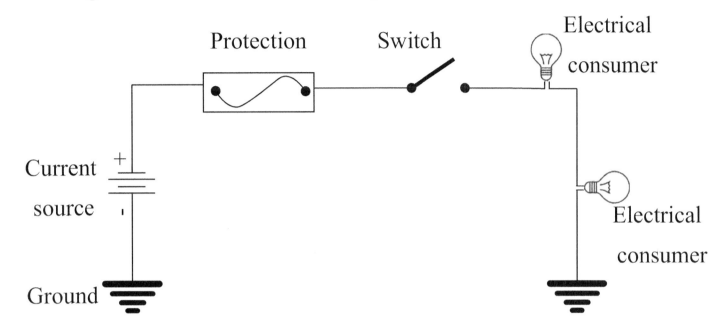

Electrical circuit in series

In series connection: The receivers are mounted one beside another. If the filament of the lamp cuts off: All the lamps do not work.

IV-II- Parallel circuit:

In a parallel circuit, the electrical consumers are arranged in a branch and provide individual paths for the current flow. Each branch has its own consumer and its own grounding. If one branch is interrupted, the other will continue operating.

Resistance in a parallel circuit:

In such an arrangement, the total resistance of the circuit is less than the resistance of any consumer on its own.

Current in a parallel circuit:

When consumers are connected in parallel, the current is divided into two or more branches. In such an arrangement, the current has more than one path to reach the ground, which implies that the current in a parallel circuit increases as the number of consumers increases, in contrast to a series circuit.

Voltage in a parallel circuit:

When consumers are connected in parallel, the voltage is the same for each branch.

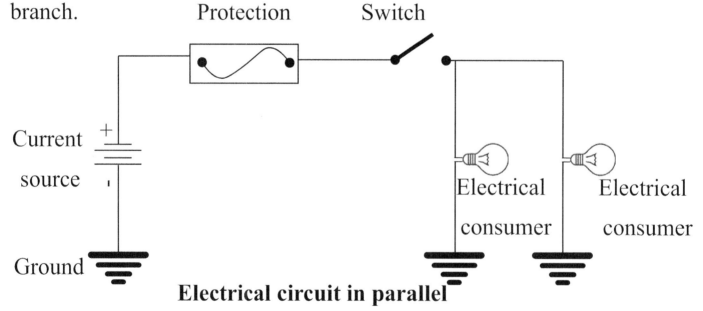

Electrical circuit in parallel

In parallel connection: The circuit is divided into as many times as there are receivers. If the filament of the lamp cuts off: The functioning of the other lamps is not modified.

This is why in a car, the receivers are mounted: in parallel.

IV-III- Series-parallel circuit:

A series-parallel circuit is a combination of series and parallel circuits.

Resistance in a series-parallel circuit:

In the branch of the circuit that contains consumers in series, the resistance is equal to the sum of the consumers' resistances, as in a series circuit. In the branch, in which the consumers are in parallel, the total resistance is less than the resistance of any consumer on its own, as in a parallel circuit.

Current in a series-parallel circuit:

The current flow will be different in each branch of the circuit, even if all consumers have the same resistance.

Voltage in a series-parallel circuit:

Either in the branch of the circuit that contains consumers in series, or in the branch in which the consumers are in parallel, the potential difference or voltage will be different in each branch of the circuit, even if all consumers have the same resistance.

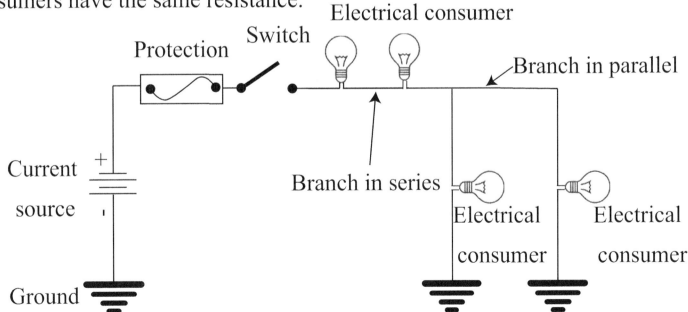

Electrical circuit in series-parallel

151

V- Protection of electrical circuits:

An electrical circuit must always have a protective device in the form of a fuse, fuse wire or circuit breaker to provide protection against current overload. Without a protective device, a short circuit would allow unlimited current to flow and the electrical wires would overheat and burn, possibly causing a fire.

V-I Fuses :

The fuse is a filament made of an amalgam of lead and tin which has the property of melting under the effect of a current overload caused by an electrical device or a ground short circuit. In such a case, the fuse overheats due to the excessive current, and melts. The circuit is then opened to prevent the breaking of the electrical device or conductor. When this happens, the circuit should be checked to find out what made the fuse melt.

Fuses are rated in amperes. In a circuit protected by a ten-amp fuse, if the current exceeds ten amperes, the fuse melts and interrupts the current flow. Fuses are mostly collected in a fuse box.

To sum up

1. Elementary Electrical Circuit = Source of electrical energy (Battery) + Conductors (Electric Wires) + Loads (Lamp) + Switch + Fuse.

2. If the current suddenly increases, the fuse must burn and not the transformers or the electrical conductors.

Calculation of electrical circuits by Ohm's Law

Objectives:

This lesson is devoted to calculations of electrical circuits by Ohm's Law.

After careful study of this lesson, you will be able to easily:

- Apply Ohm's law to a simple circuit;
- Apply Ohm's law to a complex circuit;
- Apply Ohm's Law to a direct current circuit.

I- Calculation related to a simple electrical circuit:

Consider the simple electrical circuit shown in the following figure. We want to calculate, by applying Ohm's law, the value of:

- The resistance R;
- The current I;
- The voltage of the source E.

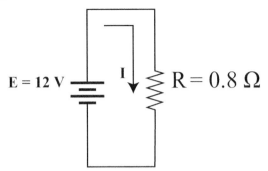

E = 12 V   R = 0.8 Ω

Current in a simple electrical circuit.

Ohm's law states that: **E = R.I** Let's draw this law in the form of a triangle as follows:

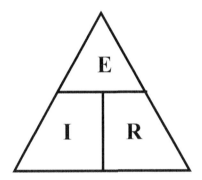

I-1- Calculation of the resistance:

We want to calculate the resistance value R so as to limit the current I to 3 amperes, knowing that the voltage of the source E is 12 volts.

Using the triangle, we can calculate the resistance R as follows:

1. Remove the resistance R box

2. The two remaining boxes give us the formula we need to apply to determine the resistance R.

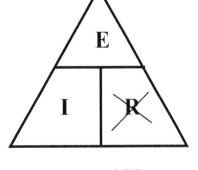

$$R = \frac{E}{I}$$

155

Numerical application:

$$R = E / I = 12\ V / 3\ A = 4\ \Omega$$

If we could have allowed a current of 4 amps in the circuit, the resistance would have been smaller. That is:

$$R = U / I = 12\ V / 4\ A = 3\ \Omega$$

I-II- Calculation of the current:

We must calculate the current intensity I, in amperes, of a current consumer, knowing that the voltage of the source E is 12 volts and that the resistance R of the consumer is 1.2 ohm.

Using the triangle, we can calculate the current intensity I as follows:

1. Remove the current intensity I box;

2. The two remaining boxes give us the formula we need to determine the current intensity I.

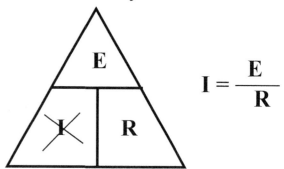

$$I = \frac{E}{R}$$

Numerical application:

$$I = U / R = 12\ V / 1.2\ \Omega = 10\ A$$

Now assume that the resistance is five times greater 6 Ω and the voltage remains the same 12 V. We then obtain:

$$I = U / R = 12\ V / 6\ \Omega = 2\ A$$

Ohm's law formula thus indicates that the current is five times smaller. But, if we keep the 1.2 ohm resistance and reduce the voltage of the current to 2.4 volts, we still get 2 amps.

$$I = U / R = 2.4 \text{ V} / 1.2 \ \Omega = 2 \text{ A}$$

Note:

To decrease the current, we only need to increase the resistance or decrease the voltage. The opposite is also true.

I-III- Calculation of the voltage:

We want to determine the voltage of the source of E. We know the resistance R= 6 ohms, and we want to make an electric current flow through it with an intensity I of 2 amps.

Still using the triangle, we can calculate the voltage of the source of E as follows:

1. Remove the source voltage of E box;

2. The two remaining boxes give us the formula we need to determine the source voltage of E.

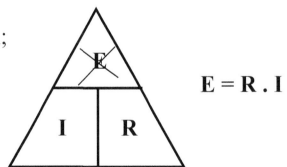

$$E = R \cdot I$$

Numerical application:

$$E = R \cdot I = 6 \ \Omega \cdot 2 \text{ A} = 12 \text{ V}$$

If the resistance were five times greater and we wanted to get the same current, a five-times greater voltage is required for the source. That is:

$$E = R . I = 30 \, \Omega . 2 \, A = 60 \, V$$

II- Calculation related to electric circuits in series:

The following figure shows the assembly diagram of an electrical circuit in series consisting
of a 12 V source and three resistors R_1, R_2 and
R_3 connected end to end of 1, 2 and 3 Ω respectively.

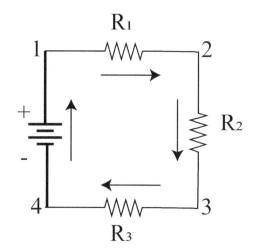

II-I- Calculation of the total resistance:

<u>Reminder</u>

To calculate the total resistance (R_{tot}) of the circuit, we sum all the values
of the resistors. Hence:

$$R_{tot} = R_1 + R_2 + R_3$$
$$R_{tot} = 1 \, \Omega + 2 \, \Omega + 3 \, \Omega$$
$$R_{tot} = 6 \, \Omega$$

II-II- Calculation of the current:

The current intensity I in a series circuit is calculated, according to Ohm's
law, as follows:

$$I = E / R_{tot} = 12 \, V / 6 \, \Omega = 2 \, A$$

In a series circuit, the same current intensity flows through all elements of the circuit. Therefore, if the current intensity in a series circuit is measured, it will be the same at any point in a series circuit.

II-III- Calculation of the voltage:

The voltage E in a series circuit is calculated, according to Ohm's law, as follows: multiplying the value of the current intensity I by the value of each resistance R in the circuit. Therefore, the voltage drop at the terminals of the three resistors is distributed as follows:

$$V_1 = R_1 . I = 1\ \Omega . 2\ A = 2\ V$$
$$V_2 = R_2 . I = 2\ \Omega . 2\ A = 4\ V$$
$$V_3 = R_3 . I = 3\ \Omega . 2\ A = 6\ V$$

Thus the total voltage drop in this circuit is:

$$V_{tot} = V_1 + V_2 + V_3 = 12\ V$$

Note:

- The sum of the voltage drops of each resistor is equal to the voltage of the source.

- The higher the resistance in a series circuit, the higher the voltage drop.

II-IV- Voltage divider (Resistor in series):

A- Formula for voltage division for two branches:

With the previous circuit, we made a voltage divider (3 V on one side and 7 V on the other),

The current of 1 mA flows through the whole circuit, it is constant at all points.

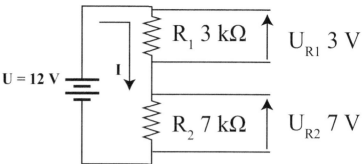

The arrow represents the current, the conventional direction (by convention) indicates that it flows from + to - when in fact physically it goes in the opposite direction. (It's historical, they made a mistake when establishing the convention, you were not born).

The voltage at the terminals of each resistor is proportional to the value of the resistor, in other words the higher the value the higher the voltage (it is called the voltage drop).

And finally...

Do I have to calculate the current to determine the voltage drop at the terminals of each resistor?

No, it is enough to calculate the ratio (the proportionality) of the voltage divider like this:

We want to calculate the voltage at the terminals of R2:

$$U_{R2} = (R_2 / (R_1 + R_2)) \times U$$

We want to calculate the voltage at the terminals of R1

$$U_{R1} = (R_1 / (R_1 + R_2)) \times U$$

To be remembered

In a voltage divider, the voltage drop at the terminals of a resistor or combination of resistors in a series circuit is equal to the ratio of the value of that resistor to the total resistance, multiplied by the source voltage.

B- General formula:

Simply apply the following formula:

$$V_X = \frac{R_X}{R_{TOT}} V_S$$

Where:

V_X is the voltage to be sought at the terminals of the resistor Rx;

R_{TOT} is the total or equivalent resistance;

V_S is the voltage supplied by the source.

Is the potentiometer an adjustable voltage divider?

We have learned that a potentiometer is a variable resistor with three terminals.

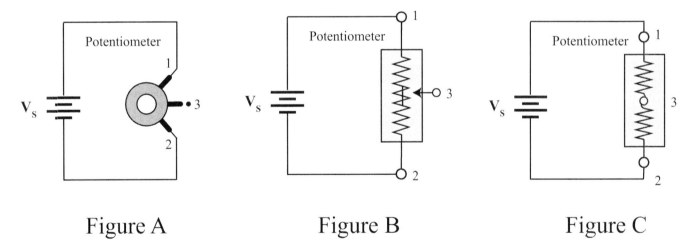

| Figure A | Figure B | Figure C |

Figure A shows a potentiometer connected to a voltage source. Note that the end terminals are designated by 1 and 2, while the wiper is designated by 3. The potentiometer acts as a voltage divider, which can be shown by separating the total resistance into two parts (Figure C). The first part is the

161

resistance between terminal 1 and terminal 3 (R13); and the second part is the resistance between terminal 3 and terminal 2 (R32). Thus, a potentiometer is a voltage divider with two resistors that can be adjusted manually. Potentiometers are used as voltage dividers in the volume control of radio or television sets. Moreover, voltage dividers can be used as a gasoline level sensor in a car tank.

III- Calculation related to electrical circuits in parallel:

While in a series circuit the electric current traverse only one way through the circuit, it flows in several branches in a circuit in parallel. The resistors are connected independently of each other. The following figure shows the assembly diagram of an electrical circuit in parallel consisting of a 12 V source and three resistors R1, R2, and R3 in parallel of 10, 5, and 2 Ω respectively.

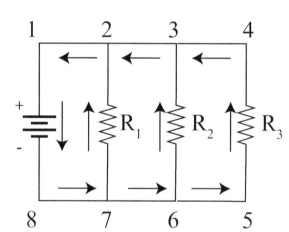

III-1- Calculation of the total resistance:

$$R_{TOT} = 1.25 \ \Omega$$

III-2- Calculation of the voltage:

We can notice that the source voltage is the same at the input of each branch of a circuit in parallel whatever the number of resistors in parallel:

A- Current division formula for two branches:

You already know how to use ohm's law to obtain the current in each branch in parallel when the voltage and resistance are known. Now, when the voltage is unknown and the total current is known, you can calculate the currents I_1 and I_2 in each branch (Next circuit) using the following formulas:

$$I_1 = (R_2 / (R_1 + R_2)) . I_T$$
$$I_2 = (R_1 / (R_1 + R_2)) . I_T$$

To be remembered:

The current in one of the two branches is equal to the resistance of the opposite branch, divided by the sum of the two resistances and multiplied by the total current.

B- General formula:

Next figure shows a parallel circuit with n branches. The current in a branch is obtained by using the formula:

$$I_n = (R_{TOT} / R_x) I_{TOT}$$

Where:

I_X is the current in any branch (I_1, I_2, ...);

R_X is the resistance in any branch (R_1, R_2, ...).

Example:

Determine the current through each resistor:

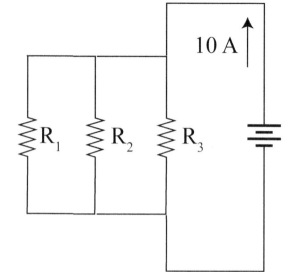

$R_1 = 680\ \Omega$

$R_1 = 330\ \Omega$

$R_1 = 220\ \Omega$

Solution:

First calculate the total parallel resistance.

$$R_{TOT} = \cfrac{1}{\cfrac{1}{R_1} + \cfrac{1}{R_2} + \cfrac{1}{R_3}} = 111\ \Omega$$

The total current is 10 A. According to the general formula of the current divider, we obtain:

$$I_1 = \frac{R_{TOT}}{R_1}\, I_{TOT} = 1.63\ A$$

$$I_2 = \frac{R_{TOT}}{R_2}\, I_{TOT} = 3.36\ A$$

$$I_3 = \frac{R_{TOT}}{R_3}\, I_{TOT} = 5.05\ A$$

To sum up:

Ohm's Law:

A series circuit is controlled by three rules which are:

- The total resistance (R_{TOT}) of the circuit is equal to the sum of all resistances present in the circuit;
- The current intensity (I) is the same at any point in the circuit;
- The sum of all voltage drops (V) in the circuit must be equal to the electromotive force of the source.

A circuit in parallel is controlled by three rules which are:

- In a parallel circuit, the total resistance (R_{TOT}) of the circuit is always less than the value of the lowest resistance in the electrical circuit;
- In a parallel circuit, the current intensity (I) is a function of the resistance of each branch of the circuit and the total resistance in the electrical circuit
- The voltage (V) in each branch of a parallel circuit is equal to the voltage of the source.

Multimeter

Objectives:

A multimeter is a measuring instrument that can measure multiple electrical properties.

After a careful study of this lesson, you will be able to easily:

- Describe a multimeter;

- Know the functions of a multimeter;

- Use a multimeter correctly.

Multimeter:

Different measuring devices are used to measure voltage, current and resistance:

- Voltmeter: to measure the voltage;

- Ammeter: to measure the current;

- Ohmmeter: to measure resistance.

These three measuring functions are combined in one measuring device called a multimeter.

A standard multimeter has different measurement capabilities:

- DCA (different measuring ranges for DC current (mA, A));

- ACA (different measuring ranges for alternating current);

- DCV (different measuring ranges for DC voltage (mV, V));

- ACV (different measuring ranges for AC voltage);

- Ω (different measuring ranges for resistance).

Other optional functions are often available, you may use them to check the status of certain electronic components such as:

- Diode test;

- Transistors test;

- Temperature;

- Continuity test (vibrator).

This device does not have a specified electrical symbol. The electrical symbol will be, according to the measurement:

- That of a voltmeter: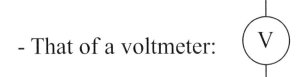

- Of an ammeter:

- Or of an ohmmeter: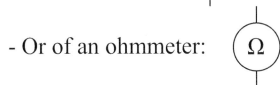

The figure below shows a drawing of a simple digital display multimeter.

You can see the screen at the top, the selector in the center (where the function switch is), and around the selector, the various functions such as;

▶ Voltmeter marked with the symbol V;

▶ Ammeter marked by the symbol A;

▶ Ohmmeters marked by the symbol Ω.

A multimeter is not enough on its own, it needs wires connected to what we want to measure. These wires consist of a touch point, electrical leads, and a plug (connector) to be attached with the multimeter.

- The RED cord for positive polarities.

- The BLACK cord for negative polarities.

- The cables and test tips must be clean and free of damage.

- The measuring cables must be correctly attached to the connection sockets used for the measuring range.

Functions of a multimeter:

Let's study the case of a simple multimeter as shown in the following figure.

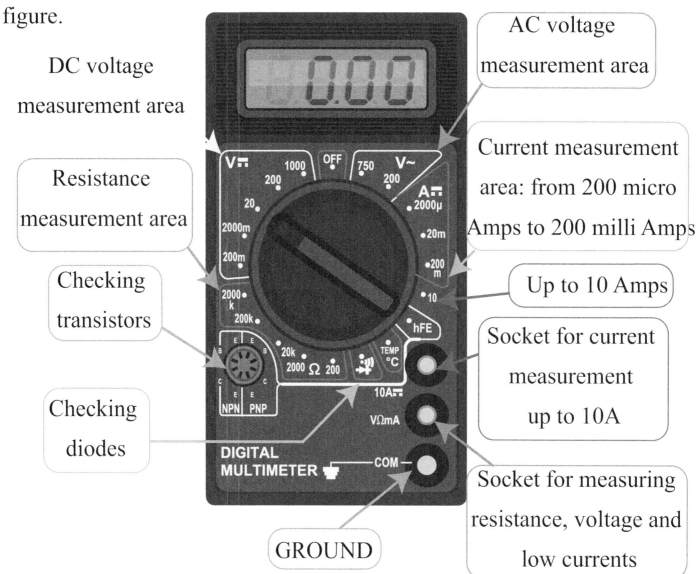

DC voltage measurement area

AC voltage measurement area

Resistance measurement area

Current measurement area: from 200 micro Amps to 200 milli Amps

Checking transistors

Up to 10 Amps

Checking diodes

Socket for current measurement up to 10A

GROUND

Socket for measuring resistance, voltage and low currents

170

Always set the multimeter to the scale value just above the expected value. For example, to measure a conventional socket, the expected value is 230 Volt, so set the multimeter to 250 Volt.

The multimeter fuses:

For its protection, the multimeter is equipped with fuses. If the ammeter function fails, check them.

Current measurement:

Measures the amount of current flowing in an electrical circuit.

THE AMPERE METER IS CONNECTED IN SERIES.

Be careful: if I > 10 A = deterioration of the fuse.

The clamp ammeter:

For currents higher than 10 A, a clamp ammeter is used.

The voltmeter must be positioned on the V DC gauge because the measurement will be displayed in mV (and not in A).

A conversion is necessary according to the gauge (1 mV / A).

Ex: During a measurement carried out using a clamp ammeter, a voltage of 100 mV corresponds to a consumption of 100 A of the controlled element.

Voltage measurement:

You can measure the voltage ...

- Between two points in a circuit;

- Between any point in a circuit and ground;

- In any component in the circuit.

Technicians are concerned with the measurement of voltages in different applications:

- Measurement of the voltage supplied by a source.

- Measurement of the voltage available in a circuit point and compares it with the one provided by the manufacturer.

- The voltage drops between the terminals of a component are compared with that provided by the manufacturer.

Measurement of a potential difference between 2 points of a circuit.

A VOLTMETER IS CONNECTED IN PARALLEL

Resistance measurement:

The ohmmeter function

It permits to measure the resistance of an element.

In ohmmeter function, the multimeter uses a battery.

The accuracy of the measurement depends on its condition. Never measure a resistance under voltage.

Key tips:

- The current always takes the easiest path.

Safety Instructions:

Improper use of the multimeter can:

- Cause personal injuries

- Damage the measuring device;

- Apply the safety instructions according to the systems diagnosed

E.g.: the use of an ohmmeter or any current generating source, on a pyrotechnic igniter, is FORBIDDEN (risk of triggering).

- Respect the gauges of the device

Ex: measurement of currents higher than 10 A.

- Select the right function for the right measurement

Ex : voltage measurement in ohmmeter function

- Connect the cables correctly according to the measurement

Ex : voltage measurement with the leads in ammeter position.

Printed in Great Britain
by Amazon

11888713R00099